The Open University

M381 Number Theory and Mathematical Logic

Mathematical Logic **Unit 6**

Formal Number Theory I

Prepared by the Course Team

The M381 Mathematical Logic Course Team

The Mathematical Logic half of the course was produced by the following team:

Roberta Cheriyan	*Course Manager*
Derek Goldrei	*Course Team Chair* and *Academic Editor*
Jeremy Gray	*History Consultant*
Mary Jones	*Critical Reader*
Roger Lowry	*Publishing Editor*
Alan Pears	*Author*
Alan Slomson	*Author*
Frances Williams	*Critical Reader*

with valuable assistance from:

The Maths Production Unit, Learning & Teaching Solutions:
Becky Browne, Jim Campbell, Nicky Kempton, Bill Norman, Sharon Powell, Katie Sayce, Penny Tee

Alison Cadle	*TEX Consultant*
Michael Goldrei	*Cover Design Consultant*
Vicki McCulloch	*Cover Designer*

The external assessor was:

Jeff Paris	*Professor of Pure Mathematics, University of Manchester*

The Course Team would like to acknowledge their reliance on the previous work of Alan Slomson and of Alex Wilkie, Professor of Mathematical Logic, University of Oxford.

This publication forms part of an Open University course. Details of this and other Open University courses can be obtained from the Student Registration and Enquiry Service, The Open University, PO Box 197, Milton Keynes, MK7 6BJ, United Kingdom: tel. +44 (0)870 300 6090, e-mail general-enquiries@open.ac.uk

Alternatively, you may visit the Open University website at http://www.open.ac.uk where you can learn more about the wide range of courses and packs offered at all levels by The Open University.

To purchase a selection of Open University course materials, visit http://www.ouw.co.uk, or contact Open University Worldwide, Michael Young Building, Walton Hall, Milton Keynes, MK7 6AA, United Kingdom, for a brochure: tel. +44 (0)1908 858793, fax +44 (0)1908 858787, e-mail ouw-customer-services@open.ac.uk

The Open University, Walton Hall, Milton Keynes, MK7 6AA.

First published 2004. Reprinted as new edition 2007, with corrections.

Edited, designed and typeset by The Open University, using the Open University TEX System.

Printed and bound in the United Kingdom by The Charlesworth Group, Wakefield.

ISBN 978 0 7492 2272 7

3.1

CONTENTS

INTRODUCTION

In this unit we complete the description of our formal system and start to discuss how to use it as a framework for proving theorems of number theory.

In *Unit 4* we introduced a formal language for number theory and in *Unit 5* we described what is meant by a formal proof. We also discussed the seven rules of proof that enable us to handle the connectives and the quantifiers of the formal language. In Section 2, we introduce the two remaining rules of proof, relating to the identity symbol, to complete the nine rules of proof of our formal system.

By requiring that each of our rules is logically valid, we ensure that whenever we have a formal proof within our system showing that the formula ψ can be derived from assumptions $\phi_1, \phi_2, \ldots, \phi_k$, then ψ is a logical consequence of $\phi_1, \phi_2, \ldots, \phi_k$, that is, ψ is true in every interpretation in which $\phi_1, \phi_2, \ldots, \phi_k$ are true. This is a consequence of the *Correctness Theorem*, which we shall discuss in Section 2. We shall also discuss a far more remarkable result, called the *Adequacy Theorem*, that our system with just these nine rules of proof is sufficiently powerful for the converse of the Correctness Theorem to hold, so that if ψ is a logical consequence of $\phi_1, \phi_2, \ldots, \phi_k$ then there is a formal proof deriving ψ from the assumptions $\phi_1, \phi_2, \ldots, \phi_k$.

The economy of setting up our formal system with just nine rules is helpful from the point of view of obtaining theoretical results about the system and means that there are fewer rules whose correct application a machine would have to check. However, it has the down side that finding formal proofs can be tougher than if we had many more rules of proof. So, in Section 1, we discuss some useful strategies for finding formal proofs, including a powerful technique called *proof by contradiction.*

This economy in drawing up the rules of proof corresponds to the economy that is normally practised in stating the axioms of algebraic structures. For example, when stating the group axioms, we need only specify the *existence* of an identity element. There is no need to stipulate in the axioms that it is *unique*, since we can deduce this fact from the axioms. Choosing a minimal set of axioms for groups reduces the amount of work needed whenever we want to check that a particular algebraic structure is a group.

The particular case of the Correctness Theorem when there are no assumptions says that if a formula ψ can be derived from no assumptions, then ψ is true in every interpretation, and hence ψ is true in the standard interpretation $\mathcal{N}$. However, we clearly cannot expect the converse to hold. There are sentences which are true in $\mathcal{N}$, for example $(\mathbf{0}' + \mathbf{0}') = \mathbf{0}''$, that we cannot expect to be derivable using no assumptions, as our rules of proof do not take into account the intended interpretation of the arithmetical symbols. Since our aim is to establish a formal system for number theory, we see that we still need to add some arithmetical axioms that encapsulate the basic properties of the operations of addition and multiplication of natural numbers. We do this in the Section 3 of this unit. We then explore, in *Unit 7*, some of the sentences of our formal language which can be derived in our formal system using these axioms as assumptions.

1 TECHNIQUES OF PROOF

In this section we look at some techniques for obtaining formal proofs using the seven rules of proof introduced in *Unit 5*.

We have already given some of these techniques in *Unit 5*. For example, if we are seeking to find a formal proof of a formula ψ from the assumptions $\phi_1, \phi_2, \ldots, \phi_k$, then a sensible starting point is to introduce the formulas $\phi_1, \phi_2, \ldots, \phi_k$ as assumptions using the Assumption Rule (Ass). We have also noted that, when an assumption has the form $\forall v\, \phi$, it is often a good move to use the Universal Quantifier Elimination Rule (UE) to remove the universal quantifier; and, when an assumption has the form $\exists v\, \phi$, it is often a good idea to introduce ϕ as an additional assumption with a view to eventually using the Existential Hypothesis Rule (EH).

The logical form of the formula that we are trying to derive also gives a useful clue as to how to proceed. We see from the formation rules for formulas that, unless the desired conclusion is an atomic formula, it will have one of the forms $\neg\phi$, $(\phi \,\&\, \psi)$, $(\phi \vee \psi)$, $(\phi \rightarrow \psi)$, $(\phi \leftrightarrow \psi)$, $\forall v\, \phi$ or $\exists v\, \phi$. In the examples in *Unit 5* we have already indicated that the Universal and Existential Quantifier Introduction Rules (UI and EI) are usually what is needed to derive formulas of the form $\forall v\, \phi$ or $\exists v\, \phi$. We also noted that, when it comes to deriving implications, that is, formulas of the form $(\phi \rightarrow \psi)$, then the Conditional Proof Rule (CP) is usually helpful.

Unit 4, Subsection 2.1.

We need to complete the picture by discussing strategies which are usually helpful when the formula we are trying to derive has one of the other forms, namely $\neg\phi$, $(\phi \,\&\, \psi)$, $(\phi \vee \psi)$ and $(\phi \leftrightarrow \psi)$. In each case it is usually a good strategy to aim to use the Tautology Rule (Taut). Each use of this rule depends on a particular formula being a tautology. Often it will not be too difficult to see which tautology to use, but in some cases it will be less obvious. An important example of this latter kind occurs when we are trying to prove a negation, that is, a formula of the form $\neg\phi$.

We look at one very powerful technique for proving negations, known as *proof by contradiction*, in Subsection 1.1. Then, in Subsection 1.2, we look at strategies which are helpful when trying to prove formulas of the form $(\phi \,\&\, \psi)$, $(\phi \vee \psi)$ and $(\phi \leftrightarrow \psi)$, and summarize the techniques for obtaining formal proofs. To help use the Tautology Rule, it is very useful to know a few standard tautologies and we give some of these in Subsection 1.3.

1.1 Proof by contradiction

A very common and effective technique in everyday mathematics for proving that some statement is not true is to show that the assumption that it is true leads to a contradiction. This method of proof is called *proof by contradiction* or *reductio ad absurdum*.

We used this method of proof in the example of an informal proof which we studied in detail at the beginning of Section 1 of *Unit 5*. In that case we wished to deduce from the assumption 'n^2 is even' the conclusion 'n is even'. We did this by assuming that 'n is odd' and from this, after some work, we were able to deduce the contradiction 'n^2 is odd and n^2 is even', or equivalently that 'n^2 is even and n^2 is not even'. From this we were able to draw the conclusion that, given the underlying assumption that 'n^2 is even', it follows that 'n is odd' is not true, that is, that 'n is even'.

This is a typical example of an informal proof by contradiction. We now look at it more closely from a formal viewpoint.

A *contradiction* is the simultaneous assertion of a statement and its negation. The formal counterpart of this is a formula of the form $(\phi \,\&\, \neg\phi)$. Thus the formal counterpart of a proof by contradiction is the claim that if from ψ and possibly other assumptions we can derive a contradiction $(\phi \,\&\, \neg\phi)$ then from the other assumptions alone we can derive $\neg\psi$. We now give the argument which justifies this claim, and this argument yields the technique which we can often use to prove formulas of the form $\neg\psi$.

Suppose then that we have a formal proof of the contradiction $(\phi \,\&\, \neg\phi)$ from the assumptions $\theta_1, \theta_2, \ldots, \theta_k$ and ψ. This means that we have a formal proof whose last line has the form

$$t_1, t_2, \ldots, t_k, s \quad (n) \quad (\phi \,\&\, \neg\phi) \quad \text{`justification'}$$

The rule of proof that has been used to obtain the last line of the formal proof is not relevant here, so we have simply written 'justification' on the right of this line.

where $t_1, t_2, \ldots, t_k, s$ are the numbers of the lines on which the formulas $\theta_1, \theta_2, \ldots, \theta_k$ and ψ have been introduced as assumptions. If we now make one use of each of the Conditional Proof and Tautology Rules, we can achieve our desired conclusion:

$$\begin{array}{rlll} t_1, t_2, \ldots, t_k, s & (n) & (\phi \,\&\, \neg\phi) & \text{`justification'} \\ t_1, t_2, \ldots, t_k & (n+1) & (\psi \rightarrow (\phi \,\&\, \neg\phi)) & \text{CP}, n \\ t_1, t_2, \ldots, t_k & (n+2) & \neg\psi & \text{Taut}, n+1 \end{array}$$

It is easily seen that we have made a correct use of the CP Rule on line $n+1$. One of the assumptions in force on line n is the formula ψ. This is the assumption introduced on line s. Using the CP Rule enables us to drop this assumption, and to introduce ψ as the antecedent of the implication on line $n+1$.

On line $n+2$ we have applied the Tautology Rule to line $n+1$. For this to be a correct use of this rule, we need to ensure that the formula $((\psi \rightarrow (\phi \,\&\, \neg\phi)) \rightarrow \neg\psi)$ is a tautology. To check this, we use a truth table:

$((\psi$	$\rightarrow$	$(\phi$	$\&$	$\neg$	$\phi))$	$\rightarrow$	$\neg$	$\psi)$
1	0	1	0	0	1	1	0	1
1	0	0	0	1	0	1	0	1
0	1	1	0	0	1	1	1	0
0	1	0	0	1	0	1	1	0
						↑ tautology		

We thus see that, having obtained a correct proof of the contradiction $(\phi \,\&\, \neg\phi)$ from the assumptions $\theta_1, \theta_2, \ldots, \theta_k$ and ψ, it takes only two more lines to extend it to a proof of $\neg\psi$ from the assumptions $\theta_1, \theta_2, \ldots, \theta_k$.

This method for proving $\neg\psi$ from the assumptions $\theta_1, \theta_2, \ldots, \theta_k$ is the method of *proof by contradiction*. It is a very useful method for proving negations, and we shall shortly illustrate this with several examples. First, however, we consider how to go about deriving the contradiction $(\phi \,\&\, \neg\phi)$ on which this method depends.

Since, clearly, the formula $((\phi \,\&\, \neg\phi) \rightarrow (\phi \,\&\, \neg\phi))$ is a tautology, to derive the contradiction $(\phi \,\&\, \neg\phi)$ it is sufficient to derive each of the formulas ϕ and $\neg\phi$ separately. For then a single use of the Tautology Rule enables us to derive $(\phi \,\&\, \neg\phi)$. That is the easy part. The hard thing is the choice of the formula ϕ which gives us one half of the contradiction. If we are aiming at deriving $\neg\psi$ from the assumptions $\theta_1, \theta_2, \ldots, \theta_k$, then ϕ needs to be chosen so that we can derive both ϕ and $\neg\phi$ from the assumptions $\theta_1, \theta_2, \ldots, \theta_n$ and ψ. There is no automatic rule for choosing the formula ϕ, given $\theta_1, \theta_2, \ldots, \theta_k, \psi$, but in many cases there are clues which lead us to a choice of ϕ for which the derivation of the contradiction $(\phi \,\&\, \neg\phi)$ from the assumptions $\theta_1, \theta_2, \ldots, \theta_k$ and ψ is not too difficult. This is best illustrated by some examples, and we now turn to these.

Example 1.1

We show that, for all formulas ϕ,

$$\exists v \neg\phi \vdash \neg\forall v\, \phi$$

Our starting point is the standard one. Since we want to derive $\neg\forall v\, \phi$ from the assumption $\exists v \neg\phi$, we introduce this latter formula as an assumption. Next, our experience from the previous unit suggests that, to derive $\neg\forall v\, \phi$ from this assumption, we should aim to deduce it from the assumption $\neg\phi$, and then use the EH Rule. To this end we introduce $\neg\phi$ as an assumption on line 2. We now turn our attention to the formula we are aiming to derive, namely $\neg\forall v\, \phi$. Since this is a negation, we try to derive it using proof by contradiction. This involves introducing $\forall v\, \phi$ as an assumption with a view to deriving a contradiction. This gives the first three lines of our formal proof as follows.

1	(1)	$\exists v \neg\phi$	Ass
2	(2)	$\neg\phi$	Ass
3	(3)	$\forall v\, \phi$	Ass

We are now seeking to derive a contradiction. Two thoughts now come together. First, since we have $\neg\phi$ on line 2, to complete the derivation of a contradiction all we need is to be able to derive ϕ on a subsequent line. Second, we note that the formula $\forall v\, \phi$ occurs on line 3, and that the UE Rule enables us to derive from it the formula ϕ. Thus it looks as though we can achieve our aim, and indeed we do this as follows.

1	(1)	$\exists v \neg\phi$	Ass
2	(2)	$\neg\phi$	Ass
3	(3)	$\forall v\, \phi$	Ass
3	(4)	ϕ	UE, 3
2, 3	(5)	$(\phi \,\&\, \neg\phi)$	Taut, 2, 4
2	(6)	$(\forall v\, \phi \rightarrow (\phi \,\&\, \neg\phi))$	CP, 5
2	(7)	$\neg\forall v\, \phi$	Taut, 6
1	(8)	$\neg\forall v\, \phi$	EH, 7

We have used the proof by contradiction method on lines 6 and 7. The use of the EH Rule on line 8 is legitimate as the variable v cannot have any free occurrences in the formula $\neg\forall v\, \phi$. ♦

Example 1.2

We show that, for all formulas ϕ and ψ,

$$\forall v\, (\psi \rightarrow \phi) \vdash (\exists v \neg\phi \rightarrow \neg\forall v\, \psi)$$

Our starting point is the standard one. Since we wish to derive $(\exists v \neg\phi \rightarrow \neg\forall v\, \psi)$ from the assumption $\forall v\, (\psi \rightarrow \phi)$, we introduce this latter formula as an assumption on line 1. Next we turn our attention to the formula $(\exists v \neg\phi \rightarrow \neg\forall v\, \psi)$ which we are aiming to derive. Since this is an implication, we adopt the standard strategy and make the assumption $\exists v \neg\phi$ with a view to deriving the conclusion $\neg\forall v\, \psi$ and then using the CP Rule. Again, as is also standard when we have an existential formula as an assumption, we also introduce as an assumption the formula without the initial existential quantifier, with a view to using the EH Rule. If we do this, the first three lines of our formal proof are as follows.

1	(1)	$\forall v\, (\psi \rightarrow \phi)$	Ass
2	(2)	$\exists v \neg\phi$	Ass
3	(3)	$\neg\phi$	Ass

Now we need to remember that our aim is to derive $\neg\forall v\,\psi$ from the assumptions 1 and 2, and then use the CP Rule to derive $(\exists v\,\neg\phi \rightarrow \neg\forall v\,\psi)$ from assumption 1 alone. Since $\neg\forall v\,\psi$ is a negation, we make the further assumption $\forall v\,\psi$ and aim to use proof by contradiction. Thus we add the line

4	(4)	$\forall v\,\psi$	Ass

to the above three lines. Now, we are trying to derive a contradiction. We note that we have the formula $\neg\phi$ on line 3, and thus we would achieve our aim of deriving a contradiction if we could also derive ϕ. There seems to be no immediate way to do this. However, another standard strategy now comes into play. Our formal proof now includes two universal formulas, on lines 1 and 4. We can use the UE Rule to drop the universal quantifiers from these formulas. We thus now have the following.

This proof exploits what will, with experience, be a typical procedure: peel off as many of the quantifiers as possible from the assumptions and hope then to be able to derive a contradiction.

1	(1)	$\forall v\,(\psi \rightarrow \phi)$	Ass
2	(2)	$\exists v\,\neg\phi$	Ass
3	(3)	$\neg\phi$	Ass
4	(4)	$\forall v\,\psi$	Ass
1	(5)	$(\psi \rightarrow \phi)$	UE, 1
4	(6)	ψ	UE, 4

We need to keep in mind that our current aim is to derive ϕ so that, together with the formula $\neg\phi$ on line 3, we have our desired contradiction.

At this stage there is no obvious way we can make progress by using any of the quantifier rules. It is therefore sensible to see if we can use the Tautology Rule to derive the formula ϕ which we are aiming for. Is this formula a tautological consequence of formulas that have already occurred in our proof? You might recognize that ϕ is a tautological consequence of the formulas ψ and $(\psi \rightarrow \phi)$. It is easy to check that the formula $((\psi \,\&\, (\psi \rightarrow \phi)) \rightarrow \phi)$ is a tautology and hence we can use the Tautology Rule to derive ϕ from ψ and $(\psi \rightarrow \phi)$.

This leaves us in a position to complete the formal proof using proof by contradiction. We do this as follows.

1	(1)	$\forall v\,(\psi \rightarrow \phi)$	Ass
2	(2)	$\exists v\,\neg\phi$	Ass
3	(3)	$\neg\phi$	Ass
4	(4)	$\forall v\,\psi$	Ass
1	(5)	$(\psi \rightarrow \phi)$	UE, 1
4	(6)	ψ	UE, 4
1, 4	(7)	ϕ	Taut, 5, 6
1, 3, 4	(8)	$(\phi \,\&\, \neg\phi)$	Taut, 3, 7
1, 3	(9)	$(\forall v\,\psi \rightarrow (\phi \,\&\, \neg\phi))$	CP, 8
1, 3	(10)	$\neg\forall v\,\psi$	Taut, 9
1, 2	(11)	$\neg\forall v\,\psi$	EH, 10
1	(12)	$(\exists v\,\neg\phi \rightarrow \neg\forall v\,\psi)$	CP, 11

We have used the proof by contradiction method on lines 9 and 10. The use of the EH Rule to derive line 11 from line 10 is legitimate since the variable v cannot have any free occurrences in either the formula $\neg\forall v\,\psi$ or the assumption other than $\neg\phi$ on which line 10 depends, namely the formula $\forall v\,(\psi \rightarrow \phi)$.

Note that, when we were deciding how to achieve the desired conclusion ϕ on line 7, it helped to be able to recognize that ϕ is a tautological consequence of the formulas ψ and $(\psi \to \phi)$. However, even if you were not sure what to do at this stage, you could have guessed that ϕ was a tautological consequence of ψ and $(\psi \to \phi)$ and you could have then confirmed this by using the standard truth table method to check that $((\psi \,\&\, (\psi \to \phi)) \to \phi)$ is indeed a tautology. ♦

To help you recognize such tautological consequences, we list some standard tautologies in Subsection 1.3.

In fact, we could have saved one line in the formal proof in Example 1.2 by noting that the two uses of the Tautology Rule on lines 7 and 8 could be compressed into one. You can check that $(((\neg\phi \,\&\, \psi) \,\&\, (\psi \to \phi)) \to (\phi \,\&\, \neg\phi))$ is a tautology. Hence after line 6 we could, using the Tautology Rule, derive $(\phi \,\&\, \neg\phi)$ from lines 3, 5 and 6. This would give line 7 as

$$1, 3, 4 \quad (7) \quad (\phi \,\&\, \neg\phi) \quad \text{Taut}, 3, 5, 6$$

and the proof could then be completed with lines 9, 10, 11 and 12 (renumbered 8, 9, 10 and 11) and with the appropriate modifications to the line references in the justification column. This corresponds to one's experience in everyday mathematics, where, once a proof of a theorem has been found, it is frequently possible to find ways to simplify it. However, as our concern here is with using standard techniques to find formal proofs, we shan't usually be bothered about trying to shorten our proofs.

Problem 1.1

(a) Show that, for all formulas ϕ,

$$\exists v\, \phi \vdash \neg\forall v\, \neg\phi$$

(b) Show that, for all formulas ϕ and ψ,

$$\forall v\, \phi,\ \exists v\, \neg\psi \vdash \neg\forall v\, (\phi \to \psi)$$

Proof by contradiction can also be used to derive formulas which are not negations. Our previous examples rely on the fact that the formula

$$((\psi \to (\phi \,\&\, \neg\phi)) \to \neg\psi)$$

is always a tautology. It is easily checked that

$$((\neg\psi \to (\phi \,\&\, \neg\phi)) \to \psi)$$

is also a tautology. It follows that if we can derive a contradiction from the assumptions $\theta_1, \theta_2, \ldots, \theta_k$ and $\neg\psi$, then by using the CP Rule and the Tautology Rule we can derive ψ from the assumptions $\theta_1, \theta_2, \ldots, \theta_k$. This method can be used to derive a formula ψ whatever the logical form of ψ, and thus is always worth a try if no other method seems to be available. The next example shows this method in use.

Example 1.3

We show that, for all formulas ϕ,

$$\neg\exists v\, \neg\phi \vdash \forall v\, \phi$$

In line with our standard strategy, we begin our formal proof by introducing the assumption $\neg\exists v\, \neg\phi$. We now aim to derive $\forall v\, \phi$. We adopt the usual strategy for deriving universal formulas, that is we aim to derive ϕ and then to use the UI Rule. How can we derive ϕ? There is no immediately obvious way to do this. In this situation, it is worth aiming to give a proof by contradiction on the lines indicated above. To do this we introduce the formula $\neg\phi$ as an assumption and try to derive a contradiction. If we adopt this strategy, we will have written the following two lines.

Note that assuming $\neg\exists v\, \neg\phi$ and $\neg\forall v\, \phi$ and trying to derive a contradiction doesn't get anywhere.

1	(1)	$\neg\exists v\, \neg\phi$	Ass
2	(2)	$\neg\phi$	Ass

We are seeking to derive a contradiction. We already have a negation on line 1 and we will thus have achieved a contradiction if we can also derive $\exists v\, \neg\phi$. It should then strike us that we can use the EI Rule to derive this formula. We can then use the proof by contradiction method. We are thus led to the following formal proof.

1	(1)	$\neg\exists v\, \neg\phi$	Ass
2	(2)	$\neg\phi$	Ass
2	(3)	$\exists v\, \neg\phi$	EI, 2
1, 2	(4)	$(\exists v\, \neg\phi\ \&\ \neg\exists v\, \neg\phi)$	Taut, 1, 3
1	(5)	$(\neg\phi \rightarrow (\exists v\, \neg\phi\ \&\ \neg\exists v\, \neg\phi))$	CP, 4
1	(6)	ϕ	Taut, 5
1	(7)	$\forall v\, \phi$	UI, 6

The use of the UI Rule to derive line 7 is correct as the variable v does not occur freely in the formula $\neg\exists v\, \neg\phi$ which is the assumption in force on line 6. ♦

Sometimes one has to be extra careful about the formula one obtains for the contradiction, especially when the next step of the proof is a use of the EH Rule, as the next example illustrates.

Example 1.4

The incorrect proof below is an attempt to show that, for all formulas ϕ and ψ,

$$\phi,\ \forall v\, (\neg\psi \vee \neg\phi) \vdash \neg\exists v\, \psi$$

where the variable v does not occur freely in ϕ.

Warning! This is not a correct proof!

1	(1)	ϕ	Ass
2	(2)	$\forall v\, (\neg\psi \vee \neg\phi)$	Ass
3	(3)	$\exists v\, \psi$	Ass
4	(4)	ψ	Ass
2	(5)	$(\neg\psi \vee \neg\phi)$	UE, 2
1, 2	(6)	$\neg\psi$	Taut, 1, 5
1, 2, 4	(7)	$(\psi\ \&\ \neg\psi)$	Taut, 4, 6
1, 2, 3	(8)	$(\psi\ \&\ \neg\psi)$	EH, 7
1, 2	(9)	$(\exists v\, \psi \rightarrow (\psi\ \&\ \neg\psi))$	CP, 8
1, 2	(10)	$\neg\exists v\, \psi$	Taut, 9

As $\neg\exists v\,\psi$ is a negation, our strategy is to add the assumption $\exists v\,\psi$ to the assumptions ϕ, $\forall v\,(\neg\psi \vee \neg\phi)$ and try to derive a contradiction. As the extra assumption is an existential formula, we also assume the formula without its existential quantifier, preparatory to a use of the EH Rule. We use the UE Rule to drop the universal quantifier from one of the other assumptions. We then derive a contradiction on line 7. Up to this point, the proof is correct. But there is a mistake on line 8. A correct application of the EH Rule to the formula on line 7, to replace assumption 4 by assumption 3, requires that the variable v does not occur freely in the formula on line 7 or in any of the assumptions, besides 4, on this line, namely assumptions 1 and 2. We are told that v doesn't occur freely in ϕ, which is assumption 1. Assumption 2 is the formula $\forall v\,(\neg\psi \vee \neg\phi)$, in which there are no free occurrences of v thanks to the initial $\forall v$. But there is no guarantee that v does not occur freely in the formula $(\psi \,\&\, \neg\psi)$, so we cannot use the EH Rule to obtain line 8.

We can, however, rescue the proof by deriving the contradiction $(\phi \,\&\, \neg\phi)$, in which the variable v does not occur freely, instead of $(\psi \,\&\, \neg\psi)$, which we can do using different tautologies for lines 6 and 7, as follows.

The condition that v does not occur freely in ϕ is vital for this proof. If the condition is dropped, it is no longer always valid to infer $\neg\exists v\,\psi$ from the assumptions ϕ and $\forall v\,(\neg\psi \vee \neg\phi)$. There is further discussion of this issue once we have stated the Correctness Theorem in Subsection 2.2.

1	(1)	ϕ	Ass
2	(2)	$\forall v\,(\neg\psi \vee \neg\phi)$	Ass
3	(3)	$\exists v\,\psi$	Ass
4	(4)	ψ	Ass
2	(5)	$(\neg\psi \vee \neg\phi)$	UE, 2
2, 4	(6)	$\neg\phi$	Taut, 4, 5
1, 2, 4	(7)	$(\phi \,\&\, \neg\phi)$	Taut, 1, 6
1, 2, 3	(8)	$(\phi \,\&\, \neg\phi)$	EH, 7
1, 2	(9)	$(\exists v\,\psi \rightarrow (\phi \,\&\, \neg\phi))$	CP, 8
1, 2	(10)	$\neg\exists v\,\psi$	Taut, 9

An alternative way to rescue our first attempt at a proof is to take its first seven lines and use the Tautology Rule with the tautology $((\psi \,\&\, \neg\psi) \rightarrow (\phi \,\&\, \neg\phi))$ to infer $(\phi \,\&\, \neg\phi)$ on line 8, depending on assumptions 1, 2 and 4. Then finish off the proof with the last three lines of the adjacent proof with line numbers suitably adjusted.

As v does not occur freely in ϕ, it doesn't occur freely in the formula $(\phi \,\&\, \neg\phi)$ on line 7 or in either of assumptions 1 and 2, so the use of the EH Rule on line 8 is now correct. The proof by contradiction method then leads to the desired result. ♦

Problem 1.2

(a) Show that, for all formulas ϕ and ψ,

$$\forall v\,(\neg\phi \rightarrow \psi),\ \neg\exists v\,\psi \vdash \forall v\,\phi$$

Hint: Introduce $\neg\phi$ as an extra assumption and aim for a contradiction.

(b) Show that, for all formulas ϕ and ψ,

$$\phi,\ \exists v\,(\psi \rightarrow \neg\phi) \vdash \neg\forall v\,\psi$$

where the variable v does not occur freely in ϕ.

1.2 Techniques for finding formal proofs

In this subsection we gather together standard techniques which are useful in finding formal proofs. Most of these techniques have already been used in earlier examples. Before we give a systematic list of these techniques, we discuss a few cases which we have not previously covered.

Disjunctions

Recall that by a disjunction we mean a formula of the form $(\phi \vee \psi)$. Our standard technique for proving formulas of this form is based on the fact that $(\phi \vee \psi)$ is a tautological consequence of the implication $(\neg\phi \rightarrow \psi)$. This follows from the fact that, for all formulas ϕ and ψ, the formula

$$((\neg\phi \rightarrow \psi) \rightarrow (\phi \vee \psi))$$

is a tautology, as is shown by the following truth table.

((¬	ϕ	→	ψ)	→	(ϕ	∨	ψ))
0	1	1	1	1	1	1	1
0	1	1	0	1	1	1	0
1	0	1	1	1	0	1	1
1	0	0	0	1	0	0	0

It follows that, if we have a formal proof of the formula $(\neg\phi \rightarrow \psi)$ from a given set of assumptions, then a single use of the Tautology Rule enables us to extend it to a formal proof of $(\phi \vee \psi)$ from the same set of assumptions.

We already have a technique which helps us to find a formal proof of an implication such as $(\neg\phi \rightarrow \psi)$. It is to add $\neg\phi$ to the list of assumptions and then aim to derive ψ. Then, by using the CP Rule, we obtain a formal proof of $(\neg\phi \rightarrow \psi)$ from the other assumptions. Thus, the observation that $(\phi \vee \psi)$ is a tautological consequence of $(\neg\phi \rightarrow \psi)$ means that we can use essentially the same strategy to derive disjunctions.

Example 1.5

We show that, for all formulas ϕ,

$$\vdash (\exists v\, \neg\phi \vee \forall v\, \phi)$$

Recall that the absence of any formulas to the left of the turnstile symbol, $\vdash$, indicates that there is a formal proof of which the last line consists of the formula $(\exists v\, \neg\phi \vee \forall v\, \phi)$ depending on no assumptions.

Using the technique described above for deriving disjunctions, we aim to derive first the implication $(\neg\exists v\, \neg\phi \rightarrow \forall v\, \phi)$. We shall be able to achieve this if we can derive $\forall v\, \phi$ from the assumption $\neg\exists v\, \neg\phi$. This is something we have already done in Example 1.3. Thus all we need to do in the present case is to add two lines to the formal proof of Example 1.3. On one line we use the CP Rule to get the required implication and then on the next line, by using the Tautology Rule, we obtain the formal proof of the disjunction. In this way we obtain the following formal proof.

1	(1)	$\neg\exists v\, \neg\phi$	Ass
2	(2)	$\neg\phi$	Ass
2	(3)	$\exists v\, \neg\phi$	EI, 2
1, 2	(4)	$(\exists v\, \neg\phi\ \&\ \neg\exists v\, \neg\phi)$	Taut, 1, 3
1	(5)	$(\neg\phi \rightarrow (\exists v\, \neg\phi\ \&\ \neg\exists v\, \neg\phi))$	CP, 4
1	(6)	ϕ	Taut, 5
1	(7)	$\forall v\, \phi$	UI, 6
	(8)	$(\neg\exists v\, \neg\phi \rightarrow \forall v\, \phi)$	CP, 7
	(9)	$(\exists v\, \neg\phi \vee \forall v\, \phi)$	Taut, 8

♦

Problem 1.3

Show that, for all formulas ϕ,

$$\vdash (\exists v\, \phi \vee \forall v\, \neg\phi)$$

Bi-implications

By a bi-implication we mean a formula of the form $(\phi \leftrightarrow \psi)$. The following truth table shows that, for all formulas ϕ and ψ, the formula

$$(((\phi \to \psi) \,\&\, (\psi \to \phi)) \to (\phi \leftrightarrow \psi))$$

is a tautology.

(((ϕ	$\to$	ψ)	&	(ψ	$\to$	ϕ))	$\to$	(ϕ	$\leftrightarrow$	ψ))
1	1	1	1	1	1	1	1	1	1	1
1	0	0	0	0	1	1	1	1	0	0
0	1	1	0	1	0	0	1	0	0	1
0	1	0	1	0	1	0	1	0	1	0

It follows that the formula $(\phi \leftrightarrow \psi)$ is a tautological consequence of the two formulas $(\phi \to \psi)$ and $(\psi \to \phi)$. Thus a standard technique for deriving an bi-implication of the form $(\phi \leftrightarrow \psi)$ is to derive each of the implications $(\phi \to \psi)$ and $(\psi \to \phi)$ separately, and then derive $(\phi \leftrightarrow \psi)$ by a single use of the Tautology Rule. Normally our technique for deriving the implication $(\phi \to \psi)$ will be to assume ϕ, aim to derive ψ and then use the CP Rule. Similarly to derive $(\psi \to \phi)$ we shall usually assume ψ, aim to derive ϕ and then use the CP Rule.

Example 1.6

We show that, for all formulas ϕ and ψ,

$$\vdash ((\neg\phi \to \psi) \leftrightarrow (\phi \vee \psi))$$

We have already noted that $(\phi \vee \psi)$ is a tautological consequence of $(\neg\phi \to \psi)$. So it requires just one use of the Tautology Rule to derive $(\phi \vee \psi)$ from $(\neg\phi \to \psi)$. It easily checked that, conversely, $(\neg\phi \to \psi)$ is a tautological consequence of $(\phi \vee \psi)$, and so we can adopt the same strategy for deriving $(\neg\phi \to \psi)$ from $(\phi \vee \psi)$. This leads to the following formal proof.

1	(1)	$(\neg\phi \to \psi)$	Ass
1	(2)	$(\phi \vee \psi)$	Taut, 1
	(3)	$((\neg\phi \to \psi) \to (\phi \vee \psi))$	CP, 2
4	(4)	$(\phi \vee \psi)$	Ass
4	(5)	$(\neg\phi \to \psi)$	Taut, 4
	(6)	$((\phi \vee \psi) \to (\neg\phi \to \psi))$	CP, 5
	(7)	$((\neg\phi \to \psi) \leftrightarrow (\phi \vee \psi))$	Taut, 3, 6

♦

Problem 1.4

Show that, for all formulas ϕ and ψ,

$$\vdash (\neg(\phi \,\&\, \psi) \leftrightarrow (\neg\phi \vee \neg\psi))$$

We are now ready to gather together the different techniques for finding formal proofs. Some of them depend on considering the logical form of the assumption formulas. Others are based on looking at the logical form of the formula that we are aiming to derive from the given assumptions.

Suppose that we are aiming to derive a conclusion ψ from a given set of assumptions $\phi_1, \phi_2, \ldots, \phi_k$.

You might find it useful when you review your work so far, for instance when revising for the exam, to see how these techniques have been used in earlier examples in this unit and in *Unit 5*.

(T1) Begin by using the Assumption Rule to introduce each of the formulas $\phi_1, \phi_2, \ldots, \phi_k$ as assumptions.

(T2) When an assumption begins with one or more universal quantifiers, use the UE Rule to eliminate these quantifiers.

(T3) When an assumption is an existential formula of the form $\exists v\, \theta$, introduce θ as an assumption with a view to a later use of the EH Rule.

(T4) When the desired conclusion ψ is a negation of the form $\neg\theta$, introduce θ as an assumption, aim to derive a contradiction, and then use the proof by contradiction method.

(T5) When the desired conclusion ψ is a conjunction of the form $(\theta \,\&\, \chi)$, aim to derive each of the formulas θ and χ separately, and then use the Tautology Rule to derive $(\theta \,\&\, \chi)$.

(T6) When the desired conclusion ψ is a disjunction of the form $(\theta \vee \chi)$, aim to derive the implication $(\neg\theta \to \chi)$, and then use the Tautology Rule to derive $(\theta \vee \chi)$.

(T7) When the desired conclusion ψ is an implication of the form $(\theta \to \chi)$, introduce θ as an assumption, aim to derive χ, and then use the CP Rule.

(T8) When the desired conclusion ψ is a bi-implication of the form $(\theta \leftrightarrow \chi)$ aim to derive each of the implications $(\theta \to \chi)$ and $(\chi \to \theta)$ separately, and then use the Tautology Rule to derive $(\theta \leftrightarrow \chi)$.

(T9) When the desired conclusion ψ is a universal formula of the form $\forall v\, \theta$, aim to derive θ, and then use the UI Rule.

(T10) When the desired conclusion ψ is an existential formula of the form $\exists v\, \theta$, aim to derive $\theta(\tau/v)$ for some term τ, and then use the EI Rule.

(T11) Remember that crucial steps in a derivation are often provided by use of the Tautology Rule.

(T12) If you are trying to derive a formula θ and nothing else seems to work, try using the proof by contradiction strategy of assuming $\neg\theta$ and aiming to derive a contradiction.

For example, in many proofs one uses techniques (T2) and (T3) to peel off quantifiers from the assumptions in order to obtain formulas without quantifiers to which the Tautology Rule is then applied, for instance to obtain a contradiction.

Example 1.7

We show that, for all formulas ϕ and ψ,

$$\exists v\, \neg\phi,\ \forall v(\phi \vee \psi) \vdash \exists v\, \psi$$

Following (T1) we begin by introducing the following two assumptions.

1	(1)	$\exists v\, \neg\phi$	Ass
2	(2)	$\forall v\, (\phi \vee \psi)$	Ass

We are guided by (T3) and (T2), respectively, for lines 3 and 4.

3	(3)	$\neg\phi$	Ass
2	(4)	$(\phi \vee \psi)$	UE, 2

Next, we turn our attention to the desired conclusion, $\exists v\, \psi$. (T10) suggests that we should aim first to derive ψ and then to use the EI Rule. Since none of the earlier techniques seems to offer any hope of further progress, we turn our attention to (T11) and look to see if we can use the Tautology Rule to derive ψ from formulas that we already have.

By thinking about the meaning of the connectives it should not be difficult to spot that ψ is a tautological consequence of the formulas $\neg\phi$ and $(\phi \vee \psi)$ on lines 3 and 4. This can be confirmed by checking that the formula $((\neg\phi \,\&\, (\phi \vee \psi)) \to \psi)$ is a tautology. We can thus add the following line.

2, 3	(5)	ψ	Taut, 3, 4

Now we can complete the proof in the way we have planned.

2, 3	(6)	$\exists v\, \psi$	EI, 5
1, 2	(7)	$\exists v\, \psi$	EH, 6

Note that the order of these last two steps cannot be reversed. An application of the EH Rule to line 5 would not be valid as, in general, ψ will contain free occurrences of the variable v.

The complete proof is as follows.

1	(1)	$\exists v\, \neg\phi$	Ass
2	(2)	$\forall v\, (\phi \vee \psi)$	Ass
3	(3)	$\neg\phi$	Ass
2	(4)	$(\phi \vee \psi)$	UE, 2
2, 3	(5)	ψ	Taut, 3, 4
2, 3	(6)	$\exists v\, \psi$	EI, 5
1, 2	(7)	$\exists v\, \psi$	EH, 6

♦

Problem 1.5

Show that, for all formulas ϕ and ψ,

$$\exists v\, (\phi \to \psi),\ \forall v\, \phi \vdash \exists v\, \psi$$

Example 1.8

We show that, for all formulas ϕ and ψ,

$$\exists v\, \phi,\ \forall v\, (\psi \to \neg\phi) \vdash \neg\forall v \psi$$

Following (T1) we begin our proof in the obvious way.

1	(1)	$\exists v\, \phi$	Ass
2	(2)	$\forall v\, (\psi \to \neg\phi)$	Ass

The next steps are in line with (T3) and (T2), respectively.

3	(3)	ϕ	Ass
2	(4)	$(\psi \to \neg\phi)$	UE, 2

Now we turn our attention to the desired conclusion, $\neg\forall v\, \psi$. Since this formula begins with the negation symbol, we shall aim to follow (T4). This leads to the following line.

5	(5)	$\forall v\, \psi$	Ass

Since this latest assumption begins with a universal quantifier, our next move is to follow (T2) and remove it.

5	(6)	ψ	UE, 5

We are aiming to derive a contradiction. We already have ϕ on line 3, so we shall obtain a contradiction if we can also derive $\neg\phi$. The Tautology Rule will almost certainly be needed, and we should be able to spot that $\neg\phi$ is a tautological consequence of the formulas on lines 4 and 6, since $((\psi \,\&\, (\psi \to \neg\phi)) \to \neg\phi)$ is a tautology. So we are able to continue as follows.

2, 5	(7)	$\neg\phi$	Taut, 4, 6
2, 3, 5	(8)	$(\phi \,\&\, \neg\phi)$	Taut, 3, 7

Having obtained a contradiction, we can apply the proof by contradiction method as planned.

2, 3	(9)	$(\forall v\, \psi \to (\phi \,\&\, \neg\phi))$	CP, 8
2, 3	(10)	$\neg\forall v\, \psi$	Taut, 9

Since the variable v has no free occurrences in either the formula $\neg\forall v\,\psi$ on line 10 or the assumption $\forall v\,(\psi \to \neg\phi)$ on line 2, we can now complete the proof by making the intended application of the EH Rule.

$1,2 \quad (11) \quad \neg\forall v\,\psi \quad \text{EH}, 10$

Thus the complete proof is as follows.

1	(1)	$\exists v\,\phi$	Ass
2	(2)	$\forall v\,(\psi \to \neg\phi)$	Ass
3	(3)	ϕ	Ass
2	(4)	$(\psi \to \neg\phi)$	UE, 2
5	(5)	$\forall v\,\psi$	Ass
5	(6)	ψ	UE, 5
$2,5$	(7)	$\neg\phi$	Taut, 4, 6
$2,3,5$	(8)	$(\phi \,\&\, \neg\phi)$	Taut, 3, 7
$2,3$	(9)	$(\forall v\,\psi \to (\phi \,\&\, \neg\phi))$	CP, 8
$2,3$	(10)	$\neg\forall v\,\psi$	Taut, 9
$1,2$	(11)	$\neg\forall v\,\psi$	EH, 10

♦

Problem 1.6

Show that, for all formulas ϕ and ψ,

$$\forall v\,\phi,\ \exists v\,(\psi \to \neg\phi) \vdash \neg\forall v\psi$$

Example 1.9

We show that, for all formulas ϕ, ψ and θ,

$$\forall v\,\theta,\ \exists v\,(\neg\phi \vee \psi) \vdash (\forall v\,\phi \to \neg\forall v\,\neg(\psi \,\&\, \theta))$$

We begin our proof in the standard way by introducing the two given assumption formulas and then using (T2) and (T3) as appropriate.

1	(1)	$\forall v\,\theta$	Ass
2	(2)	$\exists v\,(\neg\phi \vee \psi)$	Ass
1	(3)	θ	UE, 1
4	(4)	$(\neg\phi \vee \psi)$	Ass

Next we turn our attention to the formula $(\forall v\,\phi \to \neg\forall v\,\neg(\psi \,\&\, \theta))$ that we are aiming to derive as a conclusion. Since this is an implication we make use of (T7) to obtain the following line.

$5 \quad (5) \quad \forall v\,\phi \quad \text{Ass}$

Now, following (T7), we are aiming to derive $\neg\forall v\,\neg(\psi \,\&\, \theta)$. Since this begins with the negation symbol, we follow (T4) to introduce one more assumption, namely the desired conclusion with the initial negation symbol omitted.

$6 \quad (6) \quad \forall v\,\neg(\psi \,\&\, \theta) \quad \text{Ass}$

We now aim to derive a contradiction. Since the last two assumptions we have introduced begin with universal quantifiers, we follow (T2) to get the next two lines.

5	(7)	ϕ	UE, 5
6	(8)	$\neg(\psi \,\&\, \theta)$	UE, 6

How can we achieve a contradiction? There is more than one possibility here, but one way is to note that we have a negated formula $\neg(\psi \,\&\, \theta)$ on line 8 and hence we would obtain a contradiction if we could also derive $(\psi \,\&\, \theta)$, and we would be able to achieve this if only we could derive each of the formulas ψ and θ separately. We already have θ on line 3, so all we need to do to achieve our aim is to derive ψ.

Note that on line 4 we have $(\neg\phi \vee \psi)$ and on line 7 we have ϕ. The formula ψ is a tautological consequence of these formulas, since the formula $((\phi \,\&\, (\neg\phi \vee \psi)) \rightarrow \psi)$ is a tautology, as you can easily check.

You may recall that we used essentially the same tautology in Example 1.7 to derive line 5 of the proof given there.

We can thus achieve our aim of getting a contradiction as follows.

4, 5	(9)	ψ	Taut, 4, 7
1, 4, 5	(10)	$(\psi \,\&\, \theta)$	Taut, 3, 9
1, 4, 5, 6	(11)	$((\psi \,\&\, \theta) \,\&\, \neg(\psi \,\&\, \theta))$	Taut, 8, 10

We can now complete our strategy of using proof by contradiction.

1, 4, 5	(12)	$(\forall v\, \neg(\psi \,\&\, \theta) \rightarrow ((\psi \,\&\, \theta) \,\&\, \neg(\psi \,\&\, \theta)))$	CP, 11
1, 4, 5	(13)	$\neg\forall v\, \neg(\psi \,\&\, \theta)$	Taut, 12

We can now apply the CP Rule as saw we intended when we used (T7) to introduce $\forall v\, \phi$ as an assumption on line 5.

1, 4	(14)	$(\forall v\, \phi \rightarrow \neg\forall v\, \neg(\psi \,\&\, \theta))$	CP, 13

Finally we complete the proof with a use of the EH Rule, as we intended when we used (T3) to create line 4.

In this example, the last two steps in the proof could have been carried out in the reverse order. We could have applied the EH Rule to line 13 and then the CP Rule.

1, 2	(15)	$(\forall v\, \phi \rightarrow \neg\forall v\, \neg(\psi \,\&\, \theta))$	EH, 14

It should be evident that this use of the EH Rule is valid.

Thus the complete proof is as follows.

1	(1)	$\forall v\, \theta$	Ass
2	(2)	$\exists v\, (\neg\phi \vee \psi)$	Ass
1	(3)	θ	UE, 1
4	(4)	$(\neg\phi \vee \psi)$	Ass
5	(5)	$\forall v\, \phi$	Ass
6	(6)	$\forall v\, \neg(\psi \,\&\, \theta)$	Ass
5	(7)	ϕ	UE, 5
6	(8)	$\neg(\psi \,\&\, \theta)$	UE, 6
4, 5	(9)	ψ	Taut, 4, 7
1, 4, 5	(10)	$(\psi \,\&\, \theta)$	Taut, 3, 9
1, 4, 5, 6	(11)	$((\psi \,\&\, \theta) \,\&\, \neg(\psi \,\&\, \theta))$	Taut, 8, 10
1, 4, 5	(12)	$(\forall v\, \neg(\psi \,\&\, \theta) \rightarrow ((\psi \,\&\, \theta) \,\&\, \neg(\psi \,\&\, \theta)))$	CP, 11
1, 4, 5	(13)	$\neg\forall v\, \neg(\psi \,\&\, \theta)$	Taut, 12
1, 4	(14)	$(\forall v\, \phi \rightarrow \neg\forall v\, \neg(\psi \,\&\, \theta))$	CP, 13
1, 2	(15)	$(\forall v\, \phi \rightarrow \neg\forall v\, \neg(\psi \,\&\, \theta))$	EH, 14 ♦

Problem 1.7

Show that, for all formulas ϕ, ψ and θ,

$$\forall v\, (\phi \rightarrow \theta),\ \exists v\, (\psi \vee \neg\phi) \vdash (\forall v\, \phi \rightarrow \neg\forall v\, \neg(\psi \,\&\, \theta))$$

The practical question of how to spot the appropriate tautologies when seeking to find formal proofs is an important one. Most proofs involve a use of the Tautology Rule at some stage, and this requires being able to recognize when this rule can be applied. We discuss this in more detail in the next subsection, where we list some standard cases.

1.3 Some standard tautologies

We have seen that in most formal proofs we need to use the Tautology Rule at least once. Use of this rule depends on being able to spot when a formula we are aiming to derive is a tautological consequence of formulas that have already occurred in a proof. In theory this is a straightforward problem, as there is an algorithm for deciding the matter. The formula ψ is a tautological consequence of $\phi_1, \phi_2, \ldots, \phi_k$ if and only if the formula

$$((\cdots((\phi_1 \,\&\, \phi_2) \,\&\, \phi_3) \cdots \&\, \phi_k) \rightarrow \psi)$$

is a tautology, and this can be determined using the truth table method.

However, from a practical viewpoint, we don't want to spend a lot of time checking formulas which turn out not to be tautologies. So in this subsection we list a number of standard cases of tautological consequence which are frequently useful. We also indicate a method for checking whether formulas of the above form are tautologies that is often more efficient than calculating the full truth table.

We begin with a simple example to explain the notation we are going to use. We have already seen that, for all formulas ϕ and ψ, the formula $(\phi \,\&\, \psi)$ is a tautological consequence of ϕ and ψ. Thus, whenever two formulas ϕ and ψ occur in a proof, we can derive the formula $(\phi \,\&\, \psi)$ on any subsequent line. When we do this the formula $(\phi \,\&\, \psi)$ will depend on all the assumptions in force on the two lines on which the formulas ϕ and ψ occur. We indicate this by writing

$$\frac{\phi, \psi}{(\phi \,\&\, \psi)}$$

In general, we shall write

$$\frac{\phi_1, \phi_2, \ldots, \phi_k}{\psi}$$

to indicate that the formula ψ is a tautological consequence of the formulas $\phi_1, \phi_2, \ldots, \phi_k$.

Here then is our list of some standard cases of tautological consequence which are often useful. In each case, the claim that it is a tautological consequence can be justified by checking that the corresponding formula is a tautology.

1 $$\frac{\phi,\ \psi}{(\phi \,\&\, \psi)}$$
2 $$\frac{(\phi \,\&\, \psi)}{\phi}$$
3 $$\frac{(\phi \,\&\, \psi)}{\psi}$$
4 $$\frac{(\phi \vee \psi),\ \neg\phi}{\psi}$$
5 $$\frac{(\phi \vee \psi),\ \neg\psi}{\phi}$$
6 $$\frac{(\neg\phi \vee \psi),\ \phi}{\psi}$$
7 $$\frac{(\phi \vee \neg\psi),\ \psi}{\phi}$$
8 $$\frac{(\phi \rightarrow \psi),\ \phi}{\psi}$$
9 $$\frac{(\phi \rightarrow \psi),\ \neg\psi}{\neg\phi}$$
10 $$\frac{(\phi \rightarrow \psi),\ (\psi \rightarrow \theta)}{(\phi \rightarrow \theta)}$$
11 $$\frac{(\phi \leftrightarrow \psi)}{(\phi \rightarrow \psi)}$$
12 $$\frac{(\phi \leftrightarrow \psi)}{(\psi \rightarrow \phi)}$$
13 $$\frac{(\phi \rightarrow \psi),\ (\psi \rightarrow \phi)}{(\phi \leftrightarrow \psi)}$$
14 $$\frac{(\phi \rightarrow \psi)}{(\neg\psi \rightarrow \neg\phi)}$$
15 $$\frac{(\neg\phi \rightarrow \neg\psi)}{(\psi \rightarrow \phi)}$$
16 $$\frac{(\phi \rightarrow \neg\psi)}{(\psi \rightarrow \neg\phi)}$$
17 $$\frac{(\neg\phi \rightarrow \psi)}{(\neg\psi \rightarrow \phi)}$$
18 $$\frac{\neg\neg\phi}{\phi}$$
19 $$\frac{\phi}{\neg\neg\phi}$$
20 $$\frac{(\phi \,\&\, \neg\phi)}{\psi}$$

Many of the tautologies used in the logical proofs in *Unit 5* and in this unit are from this standard list. For example, in Example 1.9, standard tautological consequence 6 is used on line 9, standard tautological consequence 1 is used on lines 10 and 11, and standard tautological consequence 9 is used on line 13. You may care to look out for where these standard tautological consequences are used later in this unit and in subsequent units.

You may also care, perhaps during revision for the exam, to work out where these standard tautological consequences have been used in the logical proofs in *Unit 5* and earlier in this unit.

We have already noted that to decide whether a formula ψ is a tautological consequence of the formulas $\phi_1, \phi_2, \ldots, \phi_k$ we need to check whether the formula

$$((\cdots((\phi_1 \mathbin{\&} \phi_2) \mathbin{\&} \phi_3) \cdots \mathbin{\&} \phi_k) \rightarrow \psi) \tag{1.1}$$

is a tautology. One way to do this is to calculate the full truth table of the formula (1.1). However, a good deal of the work that this involves can be saved by taking advantage of the form of (1.1) as an implication. If any of the formulas $\phi_1, \phi_2, \ldots, \phi_k$ has the truth value 0, then, using the truth table for &, we see that the conjunction $(\cdots((\phi_1 \mathbin{\&} \phi_2) \mathbin{\&} \phi_3) \cdots \mathbin{\&} \phi_k)$ also has the truth value 0. It then follows that the whole formula (1.1) has the truth value 1, as is readily seen from the truth table for $\rightarrow$. Thus there is no need to work out in full any line of the truth table for (1.1) in which any of the formulas $\phi_1, \phi_2, \ldots, \phi_k$ has the truth value 0. Another way to put this is that, in checking to see whether (1.1) is a tautology, we need only consider those cases in which each of $\phi_1, \phi_2, \ldots, \phi_k$ has the truth value 1.

We illustrate this by returning to Example 1.9. The three successive uses of the Tautology Rule on lines 9, 10 and 11 in that example could have been replaced by a single use of this rule given by the line

The proof would then continue with lines 12 to 15 of Example 1.9, suitably renumbered.

$$1,4,5,6 \quad (9) \quad ((\psi \mathbin{\&} \theta) \mathbin{\&} \neg(\psi \mathbin{\&} \theta)) \quad \text{Taut}, 3,4,7,8$$

This is because the formula $((\psi \mathbin{\&} \theta) \mathbin{\&} \neg(\psi \mathbin{\&} \theta))$ is a tautological consequence of the formulas θ, $(\neg\phi \vee \psi)$, ϕ and $\neg(\psi \mathbin{\&} \theta)$. To justify this claim, we need to check whether the formula

$$((((\theta \mathbin{\&} (\neg\phi \vee \psi)) \mathbin{\&} \phi) \mathbin{\&} \neg(\psi \mathbin{\&} \theta)) \rightarrow ((\psi \mathbin{\&} \theta) \mathbin{\&} \neg(\psi \mathbin{\&} \theta)))$$

is a tautology. The full truth table for this formula has eight lines, but we need only check those lines in which θ, $(\neg\phi \vee \psi)$, ϕ and $\neg(\psi \mathbin{\&} \theta)$ are all true. In particular θ and ϕ must both get the truth value 1, and so we need only check two lines of the truth table, according to whether ψ is true or false. We do this now.

As the formula is so complicated, we have included the order in which we have worked out the truth values of the columns. Note that once we have discovered that all the values in the column labelled with ⑥ are 0, we know that the final values in column ⑦ must all be 1.

$((((\theta$	&	$(\neg$	ϕ	$\vee$	$\psi))$	&	$\phi)$	&	$\neg$	$(\psi$	&	$\theta))$	$\rightarrow$	$((\psi$	&	$\theta)$	&	$\neg$	$(\psi$	&	$\theta)))$
1	1	0	1	1	1	1	1	0	0	1	1	1	1	1	1	1	0	0	1	1	1
1	0	0	1	0	0	0	1	0	1	0	0	1	1	0	0	1	0	1	0	0	1
①	④	②	①	③	①	⑤	①	⑥	③	①	②	①	⑦	①	②	①	④	③	①	②	①

From column ⑦ we see that the whole formula always has the truth value 1 and hence is a tautology.

Problem 1.8

Show that, for all formulas ϕ, ψ and θ, the formula $\neg(\psi \rightarrow \theta)$ is a tautological consequence of the formulas ϕ, $(\neg\phi \vee \psi)$ and $\neg\theta$.

In our formal system, the Tautology Rule validates any deduction where one formula is a tautological consequence of other formulas. There are infinitely many different ways of using this rule, since there are infinitely many different tautologies of the form (1.1). However, in practice, we make use of a rather small set of simple special cases of this rule, such as those listed earlier as standard cases of tautological consequence.

2 THE FORMAL SYSTEM COMPLETED

In this section, we shall complete our formal system by giving rules for handling the identity symbol $=$. Plainly such rules will be vital for proving statements of number theory, which are fundamentally built up using the $=$ symbol. We shall also discuss two very important results about the formal system, the Correctness Theorem and the Adequacy Theorem, which connect what can be derived within the system with logical consequence.

2.1 Identity rules

In *Unit 5* we introduced the rules of proof of our formal system which deal with the connectives and the quantifiers. We now complete the formal system by describing two rules relating to the identity symbol: an introduction rule and a substitution rule.

The introduction rule is very straightforward. It is based on the principle that anything is identical to itself.

> ***Definition 2.1 Identity Introduction Rule (II)***
>
> For each term τ, the formula
>
> $$\tau = \tau$$
>
> may be written down on any line of a formal proof, and does not depend on any assumptions.

The rule is plainly logically valid and it is easy to see that a machine could check whether a formula has this form.

Example 2.1

We show that

$$\vdash \forall v\, v = v$$

The following formal proof establishes this claim.

(1)	$v = v$	II
(2)	$\forall v\, v = v$	UI, 1

Note that since line 1 does not depend on any assumptions, the use of the UI Rule on line 2 is legitimate since, vacuously, the variable v does not occur freely in any of the assumptions on which line 1 depends. ♦

As we can use the Identity Introduction Rule on any line of a formal proof, it shares with the Assumption Rule the property that it can be used on the *first* line of a proof. None of the other rules of our formal system has this property. So each formal proof must start with either a use of the Assumption Rule or a use of the Identity Introduction Rule.

The second rule for the identity symbol corresponds to the principle that 'equals can be substituted for equals'. That is, if two terms τ_1 and τ_2 have been shown to be equal in the sense that we have derived the formula $\tau_1 = \tau_2$, and, for some formula ϕ, we have also derived the formula $\phi(\tau_1/v)$, then this principle suggests that we should be able to substitute τ_2 for τ_1 and hence derive the formula $\phi(\tau_2/v)$.

There is, however, a technical difficulty, in terms of logical validity, in transferring this principle into a rule of our formal system. This is because there may be variables in τ_1 or τ_2 that become bound when one of these terms is substituted for the variable v in ϕ. Here is an example of what can happen.

We met a similar technical difficulty when introducing the UE Rule in *Unit 5*.

Let ϕ be the formula

$$\exists y\, y = (x + \mathbf{0}')$$

and let τ_1 and τ_2 be the terms $\mathbf{0}''$ and $(y + \mathbf{0}')$, respectively. Then $\tau_1 = \tau_2$ is the formula

$$\mathbf{0}'' = (y + \mathbf{0}') \tag{2.1}$$

$\phi(\tau_1/x)$ is the formula

$$\exists y\, y = (\mathbf{0}'' + \mathbf{0}') \tag{2.2}$$

and $\phi(\tau_2/x)$ is the formula

$$\exists y\, y = ((y + \mathbf{0}') + \mathbf{0}') \tag{2.3}$$

Formula (2.1) is true in the standard interpretation $\mathscr{N}$ if we interpret the variable y as referring to the number 1. Formula (2.2) is also true in $\mathscr{N}$, but formula (2.3) is not true in this interpretation. What has gone wrong is that the variable y in the term $(y + \mathbf{0}')$ becomes bound in formula (2.3).

To avoid this difficulty, we need to ensure that τ_1 and τ_2 are freely substitutable in ϕ, so that no variables become bound.

Definition 2.2 Substitution Rule (Sub)

Suppose τ_1 and τ_2 are terms which may be freely substituted for the variable v in the formula ϕ. Let ϕ_1 be a formula which results when some (but not necessarily all) of the free occurrences of v in ϕ are replaced by τ_1, and let ϕ_2 be the formula which results when these same occurrences of v are replaced by τ_2. Then, if on two lines of a formal proof the formulas $\tau_1 = \tau_2$ and ϕ_1 occur, on any subsequent line we may introduce ϕ_2, which will depend on all the assumptions in force on the lines on which the formulas $\tau_1 = \tau_2$ and ϕ_1 occur.

This rule can be shown to be logically valid. It is easy to see that a machine could check whether the rule has been applied correctly.

We now give some examples of formal proofs which make use of the identity rules.

Example 2.2

We show that

$$\vdash \forall x\, \forall y\, (x = y \to x' = y')$$

1	(1)	$x = y$	Ass
	(2)	$x' = x'$	II
1	(3)	$x' = y'$	Sub, 1, 2
	(4)	$(x = y \to x' = y')$	CP, 3
	(5)	$\forall y\, (x = y \to x' = y')$	UI, 4
	(6)	$\forall x\, \forall y\, (x = y \to x' = y')$	UI, 5

In terms of the notation used in stating the Substitution Rule, we have the following. The formula ϕ is $x' = x'$, τ_1 is the term x and τ_2 is the term y. Thus the formula on line 1 is $\tau_1 = \tau_2$. The formula on line 2 is ϕ_1. It may be regarded as having been obtained from ϕ by replacing the second occurrence of the variable x by the variable x. Then ϕ_2, the formula on line 3, is obtained from ϕ by replacing the second occurrence of x by an occurrence of y.

More briefly, the use of the Substitution Rule is justified since the formula on line 3 is obtained from the formula on line 2 by substituting the variable y for the second occurrence of x, and $x = y$ is the formula on line 1. ♦

Example 2.3

We show that

$$\vdash \forall x\, \forall y\, (x = y \to y = x)$$

1	(1)	$x = y$	Ass
	(2)	$x = x$	II
1	(3)	$y = x$	Sub, 1, 2
	(4)	$(x = y \to y = x)$	CP, 3
	(5)	$\forall y\, (x = y \to y = x)$	UI, 4
	(6)	$\forall x\, \forall y\, (x = y \to y = x)$	UI, 5

The use of the Substitution Rule on line 3 is justified since the formula on this line is obtained from the formula on line 2 by substituting the variable y for the first occurrence of x, and $x = y$ is the formula on line 1. ♦

Problem 2.1

Establish the following.

(a) $\vdash \forall x\, \forall y\, \forall z\, (x = y \to (z + x) = (z + y))$

(b) $\vdash \forall x\, \forall y\, \forall z\, (x = y \to (y = z \to x = z))$

There is a feature of the Substitution Rule that we need to be careful about, and which is alluded to in the solution to Problem 2.1(b). The order of the terms τ_1 and τ_2 in the statement of the rule in Definition 2.2 is critical for a correct use of the rule. Put somewhat loosely, given $\tau_1 = \tau_2$ the rule allows us to substitute τ_2 for τ_1, but it does not permit the substitution of τ_1 for τ_2. The following will make this distinction more precise. The first three lines of the proof of Example 2.2 are as follows.

1	(1)	$x = y$	Ass
	(2)	$x' = x'$	II
1	(3)	$x' = y'$	Sub, 1, 2

These three lines constitute a formal proof which shows that $x = y \vdash x' = y'$. However, to show that $y = x \vdash x' = y'$, which is indeed true, it is *not* enough to replace the formula $x = y$ on line 1 by $y = x$. This would give the following 'proof'.

Warning! This is not a correct proof!

1	(1)	$y = x$	Ass
	(2)	$x' = x'$	II
1	(3)	$x' = y'$	Sub, 1, 2

The use of the Substitution Rule on line 3 is not correct. The formula on line 3 has been obtained from that on line 2 by substituting y for the second occurrence of x, and it is claimed that this substitution is justified by the occurrence of the formula $y = x$ on line 1. However, this is not a substitution which the Substitution Rule allows. If we apply this rule to the formula $y = x$, we are only allowed to use it to substitute x for occurrences of y but not the other way round.

We ask you to give a correct formal proof to show that $y = x \vdash x' = y'$ in Problem 2.2(a).

Since a rule which also permitted substitutions in the other direction would also be logically valid, you may wonder why we formulated the Substitution Rule to permit substitutions just in one direction. It is a matter of economy in setting up our formal system, a point we remarked on in the Introduction to this unit. Substitutions in the other direction are easily catered for using our Substitution Rule since, if we are given the formula $\tau_1 = \tau_2$ and want to substitute τ_1 for τ_2, rather than τ_2 for τ_1, in some other formula, we can always first derive the formula $\tau_2 = \tau_1$, by an adaptation of the first three lines of the proof in Example 2.3, as follows.

$$
\begin{array}{llll}
1 & (1) & \tau_1 = \tau_2 & \text{Ass} \\
 & (2) & \tau_1 = \tau_1 & \text{II} \\
1 & (3) & \tau_2 = \tau_1 & \text{Sub}, 1, 2
\end{array}
$$

We can then use the Substitution Rule to substitute τ_1 for τ_2 in some other formula, provided that all the requirements of the rule are satisfied.

With care, the restrictions of the Substitution Rule need not be too great, as the following example shows.

Example 2.4

We show that

$$x = y \vdash ((x + y) + x) = ((y + x) + y)$$

Of course, we start the formal proof in the standard way by introducing the formula $x = y$ as an assumption. The next step is to find a term τ such that, by substituting y for some of the occurrences of x in the formula $\tau = \tau$, we obtain the formula $((x + y) + x) = ((y + x) + y)$ at which we are aiming. If we achieve this, we can use the II Rule to introduce the formula $\tau = \tau$ into our proof, and we can then achieve our desired conclusion by a single use of the Substitution Rule. We hope that you will agree that we have made the optimum choice of the term τ.

$$
\begin{array}{llll}
1 & (1) & x = y & \text{Ass} \\
 & (2) & ((x + x) + x) = ((x + x) + x) & \text{II} \\
1 & (3) & ((x + y) + x) = ((y + x) + y) & \text{Sub}, 1, 2
\end{array}
$$

The formula on line 3 has been obtained from that on line 2 by substituting y for the second, fourth and sixth occurrences of x. This is a legitimate use of the Substitution Rule since we have the formula $x = y$ on line 1. ♦

Plainly, formal proofs of theorems about number theory are going to involve more complicated uses of the Substitution Rule than those we have seen in this subsection. We shall see such uses in *Unit 7*, when we discuss formal theorems in the system Q, the axioms of which we shall introduce in Subsection 3.2.

Problem 2.2

Establish the following.

(a) $y = x \vdash x' = y'$

(b) $x = y \vdash ((y \cdot x) \cdot y) = ((x \cdot y) \cdot x)$

2.2 Correctness and adequacy

You have now met all nine rules of proof of our formal system: the Assumption Rule (Ass), the Tautology Rule (Taut) and the Conditional Proof Rule (CP), introduced in Section 1 of *Unit 5*; the Universal Quantifier Elimination Rule (UE) and the Universal Quantifier Introduction Rule (UI) introduced in Section 2 of *Unit 5*; the Existential Quantifier Introduction Rule (EI) and the Existential Hypothesis Rule (EH) introduced in Section 3 of *Unit 5*; and the Identity Introduction Rule (II) and the Substitution Rule (Sub), introduced in the previous subsection. The nine rules of proof, together with the formal language and its rules discussed in *Unit 4*, complete our description of our formal system. So it is now time to discuss some important results about the system.

Recall that, in Section 1 of *Unit 5*, we said that the formal proofs of our formal system should be machine-checkable and logically valid.

For a formal proof to be machine-checkable, it is sufficient that each step of the proof be machine-checkable. This will be the case provided that, for each rule of proof, there is an algorithm for determining whether it has been applied correctly. We hope that our informal discussions of this issue in *Unit 5* and in this unit have convinced you that such algorithms exist. We shall not go into any further detail here.

A full proof that the proofs of our formal system are machine-checkable would involve translating this requirement into the language of recursive functions, and checking that the relevant functions are recursive, in an analogous way to that in which we discussed URM computations in *Unit 3*.

We do, however, need to look at the logical validity of formal proofs in more detail. Recall that a formal proof of a formula ψ from assumptions $\phi_1, \phi_2, \ldots, \phi_k$ is logically valid provided that ψ is a logical consequence of the formulas $\phi_1, \phi_2, \ldots, \phi_k$, in other words provided that ψ is true in every interpretation in which each of the formulas $\phi_1, \phi_2, \ldots, \phi_k$ is true. Implicit in this definition is the fact that the set of assumptions is finite. Later, however, when investigating possible sets of axioms for number theory, we shall want to consider infinite sets of assumptions. To this end, we need to amend our definition of logical consequence, and for good measure to provide a formal definition of *derivation.*

Recall that, in the Introduction to *Unit 5*, we informally equated the idea of derivation with that of formal proof.

Definition 2.3 Derivation and logical consequence

Let S be a set of formulas and ψ a formula.

The set S may be infinite.

We say that ψ is *derivable* from S if there is a formal proof of ψ from assumptions which are all in the set S. Such a proof is called a *derivation* of ψ from S.

Note that this covers the case where we have a formal proof of ψ from no assumptions.

We say that ψ is a *logical consequence* of S if ψ is true in every interpretation in which all the formulas of S are true.

A key point to note is that any formal proof is finitely long, so that, even if S is infinite, a proof of ψ from S uses only finitely many assumptions from S. There is a further, subtle, issue in the case when the set S is infinite: is there an algorithm for checking whether a given derivation of ψ is a derivation from S? Since our formal system is machine-checkable, we know that there is an algorithm which will check whether the derivation is correct and which will identify the assumptions on which ψ depends. But if S is infinite, there might not be an algorithm for checking whether these assumptions are in S — this issue will underly much of our subsequent work on using the formal system for number theory.

When S is finite, there is no problem. The formulas in S can be put in a finite list and any assumption can be compared with each formula in the list in finitely many steps.

In light of our new, extended, definition of logical consequence, we should now state that, in order to show that the formal proofs of our system are logically valid, we need to show that, if ψ is derivable from assumptions in a set S, then ψ is a logical consequence of S, that is, ψ is true in every interpretation in which all the formulas of S are true. As before, the method for doing this is to show that, in a formal proof, the formula which appears on a given line is always a logical consequence of the assumption formulas on which that line depends. However, we don't need to check each formal proof individually: as was discussed in *Unit 5*, it is sufficient to check that each of the nine rules of proof of our formal system is logically valid, in other words that whenever a rule is applied to earlier lines on which the formula is a logical consequence of the assumptions, this remains true for the line of the proof obtained using the rule of proof in question. We have already checked this, in *Unit 5*, for the Assumption, Tautology and Conditional Proof Rules. As we have indicated in *Unit 5* and in this unit, it is possible to check that all the other rules of proof also have this property, though, since the arguments generally involve a number of not very illuminating technical details relating to free occurrences of variables, we have omitted them: we hope that our informal discussions of these rules make it plausible that they are logically valid. Taking this on trust, we are able to state the following theorem.

For the details of the arguments, consult the Suggestions for Further Reading in *Unit 8*.

Theorem 2.1 *Correctness Theorem*

If S is a set of formulas and ψ a formula such that ψ is derivable from S, then ψ is a logical consequence of S.

In particular, it follows from this theorem that if ψ is a formula which is provable from no assumptions, that is $\vdash \psi$, then ψ is true in every interpretation of our formal language.

Problem 2.3

Show that it is not the case that, for all formulas ϕ,

$$\vdash (\forall x\, \exists y\, \phi \rightarrow \exists y\, \forall x\, \phi)$$

The Correctness Theorem tells us that if ψ is *not* a logical consequence of S then there cannot be a formal proof of ψ from assumptions in S. Thus, if we can spot that ψ is not a logical consequence of S, then we know that there is no point in trying to derive ψ from S. For example, suppose we want to try to prove $\neg\exists v\, \theta$ from the assumptions ϕ and $\forall v\, (\neg\theta \vee \neg\phi)$. To see that there is no point in trying to do this, suppose that θ is the formula $v = \mathbf{0}$ and that ϕ is the formula $\neg\, v = \mathbf{0}$. The formula $\forall v\, (\neg\theta \vee \neg\phi)$ becomes $\forall v\, (\neg\, v = \mathbf{0} \vee \neg\neg\, v = \mathbf{0})$, which is true in all interpretations. However, in the standard interpretation $\mathscr{N}$, if we give v the value 3, say, then both ϕ and $\forall v\, (\neg\theta \vee \neg\phi)$ are true but $\neg\exists v\, \theta$, which is the formula $\neg\exists v\, v = \mathbf{0}$, is false. Thus $\neg\exists v\, v = \mathbf{0}$ is not a logical consequence of ϕ and $\forall v\, (\neg\theta \vee \neg\phi)$ and so there would be no point in trying to derive $\neg\exists v\, v = \mathbf{0}$ from these assumptions, as the Correctness Theorem tells us that no such formal proof can exist.

In this and other examples, it is sometimes possible to ensure that ψ is a logical consequence of S by restricting the free occurrences of certain variables. In this example, if v does not occur freely in ϕ, then $\neg\exists v\, \theta$ is a logical consequence of ϕ and $\forall v\, (\neg\theta \vee \neg\phi)$, as we saw in Example 1.4.

It is natural to ask if the converse of the Correctness Theorem is true. If the formula ψ is a logical consequence of the set of formulas S, does it follow that ψ is derivable from S? That is, will there necessarily be a formal proof of ψ from finitely many assumptions $\phi_1, \phi_2, \ldots, \phi_k$ in S? This is a deep question which amounts to asking whether our formal system is adequate for deriving all logical consequences that can be derived within our system. Indeed, more than 50 years elapsed between the first formulation of a system of formal logic similar to that presented here (published by Frege in 1879) and the proof that the answer to our question is 'yes' (published by Gödel in 1930). In fact, it took some time for the question about the adequacy of the system even to be asked!

Because of the difficulty of the proof, we content ourselves just with stating the theorem which gives the answer to our question.

For a proof, consult the Suggestions for Further Reading in *Unit 8*.

Theorem 2.2 *Adequacy Theorem*

If S is a set of formulas and ψ a formula such that ψ is a logical consequence of S, then ψ is derivable from S.

At this point we need to remind ourselves that the main focus of our interest is number theory. We seek to develop the formal system so it can prove results of number theory. At the moment we are unable to give formal proofs of such obviously true statements of number theory as

$$\forall x\,(x + \mathbf{0}') = x'$$

But this is not surprising. None of our rules makes any reference to the standard interpretation of the arithmetical symbols. Indeed we have made sure that our rules are logically valid, so that the formulas we can prove without any assumptions are those which are true in all interpretations. The formula above, though true in the standard interpretation $\mathscr{N}$, is not true in every interpretation. Hence it is not possible to derive it in our system without using any assumptions. So we need to add some 'axioms for arithmetic' which express the basic properties of the arithmetical symbols. We begin this task in the next section.

3 AXIOMS FOR NUMBER THEORY

In this section we shall explain how to use our formal system to prove some non-trivial statements of number theory. This requires us to make precise what we mean by a theory and by the axioms and theorems of a theory, which we do in Subsection 3.1. Subsection 3.2 then gives some axioms for a fragment of number theory.

3.1 Theories and theorems

In everyday mathematics the term 'theory' is frequently used, but is not precisely defined. We speak, for example, of 'the theory of groups', 'chaos theory' or, indeed, 'number theory'. In our work in mathematical logic, 'theory' has a precise definition.

Our definition of a theory depends on the notion of a *sentence*, by which, as you may recall, we mean a formula of the formal language that contains no free occurrences of variables. The lack of free occurrences of variables in sentences makes them very useful since this property ensures that, in any given interpretation of the formal language, a sentence Φ is either always true or always false: there are no free variables in Φ for which we need to provide values in order to determine the truth or falsehood of the formula. Furthermore, we can use sentences in formal proofs without having to worry about whether they satisfy the conditions on free variables that apply in some of the rules of proof. These attractive properties explain why we base our definition of a theory upon them.

We adopt the practice of using capital Greek letters for sentences.

Definition 3.1 Theory, axioms, theorems

Let S be a set of sentences of the formal language. The set of all sentences which are derivable in the formal system from assumptions in the set S is called the *theory of* S and written as $\mathrm{Th}(S)$. S is said to be a set of *axioms* of the theory.

A set of sentences T is said to be a *theory* if it is $\mathrm{Th}(S)$ for some set of sentences S. The sentences which are in T are called *theorems* of T. If a sentence Φ is a theorem of T, we indicate this by writing

$$\vdash_T \Phi$$

A theory can have several different sets of axioms. See Example 3.1.

In some books, the sentences in T are called theorems of the set S of axioms.

It follows from the Correctness and Adequacy Theorems that a sentence Ψ is derivable from S if and only if Ψ is a logical consequence of S, so we could equally well define the theory $\text{Th}(S)$ as the set of all sentences which are logical consequences of S. It is sometimes more convenient to establish results about theories using logical consequence rather than derivability within the formal system. For instance, any theory T is closed under derivation, as we show using logical consequence in the following theorem.

Ψ is the Greek letter 'capital psi'.

Theorem 3.1 Closure under derivation

Let T be a theory. Then, for all sentences Ψ, if Ψ is derivable from T then $\Psi \in T$.

Proof

Let the theory T have as axioms the set of sentences S, so that $T = \text{Th}(S)$. Suppose Ψ is a sentence derivable from T. For Ψ to be in T, we need to show that Ψ is derivable from S.

As $T = \text{Th}(S)$ consists of all sentences which are logical consequences of S, then, for any interpretation in which all sentences of S are true, all sentences of T are true. As Ψ is derivable from T, then, by the Correctness Theorem, Ψ is a logical consequence of T, so that in this interpretation Ψ is true. Thus Ψ is a logical consequence of S and, by the Adequacy Theorem, derivable from S, so is in $\text{Th}(S) = T$. ■

If you would like to try your hand at proving Theorem 3.1 directly using the formal system, rather than going via the Correctness and Adequacy Theorems, look at Additional Exercise 2 for Section 3.

Problem 3.1

Let S be a set of sentences. Show that every sentence Ψ in S is in the theory $\text{Th}(S)$.

It is useful to have a special name for an interpretation in which all the theorems of a theory are true.

Definition 3.2 Interpretation of a theory

An *interpretation* of a theory T is an interpretation of the formal language in which all the theorems of the theory are true.

Part of the proof of Theorem 3.1 shows that, if S is a set of sentences, then any interpretation of the formal language in which all the sentences of S are true is also an interpretation of the theory $\text{Th}(S)$. This can be rephrased as the following theorem.

Theorem 3.2

If T is a theory with axioms S, then any interpretation of the formal language in which all the sentences of S are true is also an interpretation of T.

Example 3.1

If the set S is empty, then $\mathrm{Th}(S)$ consists of those sentences Ψ such that $\vdash \Psi$, that is, such that Ψ is derivable from no assumptions. $\mathrm{Th}(S)$ is not empty: for example, it follows from Example 2.1 that the sentence $\forall x\, x = x$ is in $\mathrm{Th}(S)$.

Equivalently $\mathrm{Th}(S)$ consists of all sentences true in all interpretations of the formal language, and can be described as the set of theorems of quantifier logic (i.e. obtained solely from the nine rules of proof, without any special assumptions).

If S' is the set consisting of just the one formula $\forall x\, x = x$, the theory $\mathrm{Th}(S')$ is equal to $\mathrm{Th}(S)$. It is perhaps easier to explain this in terms of logical consequence than in terms of the formal proof system. Suppose that the sentence Ψ is in $\mathrm{Th}(S')$, which means that Ψ is a logical consequence of $\forall x\, x = x$. As $\forall x\, x = x$ is true in all interpretations of the formal language, it follows that Ψ is also true in all interpretations and is thus in $\mathrm{Th}(S)$. Conversely, any sentence in $\mathrm{Th}(S)$ is a logical consequence of the empty set of sentences, and thus of any other set of sentences, in particular of S', and is thus in $\mathrm{Th}(S')$. ♦

Example 3.1 illustrates that different sets of axioms, in this particular case the empty set and the set consisting of the sentence $\forall x\, x = x$, can give rise to the same theory.

Theorem 3.3

Let $\mathscr{M}$ be an interpretation of the formal language. Let T be the set of all sentences which are true in $\mathscr{M}$. Then T is a theory.

Proof

To show that T is a theory, we need to identify a set of sentences S for which $T = \mathrm{Th}(S)$. In this case, one suitable set S is T itself and we shall show that $T = \mathrm{Th}(T)$.

By Problem 3.1, every sentence in T is in $\mathrm{Th}(T)$. To show the converse, suppose that the sentence Ψ is derivable from T. Then it follows from the Correctness Theorem that Ψ is a logical consequence of T. Hence, as all the sentences in T are true in the given interpretation, so also is Ψ. Hence $\Psi \in T$. Thus $T = \mathrm{Th}(T)$, which shows that T is a theory. ■

One instance of Theorem 3.3 is of particular importance. This is the case where the interpretation in question is the standard interpretation $\mathscr{N}$. The sentences true in $\mathscr{N}$ correspond to what in everyday mathematics we would call the truths of number theory. We give this theory a special name.

Definition 3.3 Complete Arithmetic

Complete Arithmetic is the theory that consists of all those sentences of the formal language that are true in the standard interpretation $\mathscr{N}$. We shall use the abbreviation CA for this set of sentences.

The reason why we use the word 'complete' in this name will emerge shortly.

Theories like Complete Arithmetic which consist of all sentences that are true in some particular interpretation have two important properties. Let T be such a theory. Given any particular interpretation of the formal language and any sentence Φ, then precisely one of the two sentences Φ and $\neg\Phi$ is true in the given interpretation. Thus one important property of T is that either $\vdash_T \Phi$ or $\vdash_T \neg\Phi$, and the other is that we cannot have both $\vdash_T \Phi$ and $\vdash_T \neg\Phi$.

We now give two general definitions corresponding to these two properties, starting with the second.

Definition 3.4 Consistency

A theory T is said to be *consistent* if there is no sentence Φ of the formal language such that both $\vdash_T \Phi$ and $\vdash_T \neg\Phi$.

If a theory T is not consistent, so that there is some sentence Φ such that both $\vdash_T \Phi$ and $\vdash_T \neg\Phi$, then T has no interpretations, as no sentence can be both true and false in an interpretation. That means that *a theory which has an interpretation is consistent.* In particular, Complete Arithmetic is consistent.

It can be shown that the converse is also true: every consistent theory has an interpretation.

Problem 3.2

Prove that if T is a theory which is *not* consistent then every sentence of the formal language is a theorem of T, that is, show that for each sentence Ψ we have $\vdash_T \Psi$. *Hint:* Recall, from Subsection 1.3, that, for all sentences Φ and Ψ, the sentence $((\Phi \mathbin{\&} \neg\Phi) \rightarrow \Psi)$ is a tautology. Use this to show how to construct a derivation of Ψ.

Our second general definition is as follows.

Definition 3.5 Completeness

A theory T is said to be *complete* if, for every sentence Φ of the formal language, either $\vdash_T \Phi$ or $\vdash_T \neg\Phi$.

Problem 3.3

Prove that if T is a theory which is both complete and consistent then, for every sentence Φ of the formal language, precisely one of $\vdash_T \Phi$ and $\vdash_T \neg\Phi$ holds.

From the discussion preceding Definitions 3.4 and 3.5, we see that Complete Arithmetic (CA) is a complete and consistent theory.

The fact that it is a complete theory explains its name.

It should be noted that the consistency of the theory CA has no bearing on Hilbert's Question, which we mentioned at the beginning of *Unit 1* and again in the Introduction to *Unit 4*. Hilbert asked whether the consistency of number theory can be proved using only non-dubious principles of *finitary* reasoning. An argument for the consistency of number theory which is based on the existence of the standard interpretation, which is an infinite mathematical object, certainly does not meet Hilbert's requirement of a consistency proof which only uses finitary reasoning.

We shall have to wait until *Unit 8* for an answer to Hilbert's Question.

Leibniz's Question, which we mentioned along with Hilbert's Question, asks if there is an algorithm for deciding which statements of number theory are true. This question can now be reformulated in terms of the theory CA as follows:

We answer Leibniz's Question in *Unit 8* too.

> Is there an algorithm for deciding which sentences of our formal language are in the set CA?

We still cannot answer this question. However, to help us answer it later, you should find it plausible that we should search for a *finite* set of axioms for the theory CA. We begin the search for such a set of axioms in the next subsection.

We already have an *infinite* set of axioms for the theory CA, namely CA itself. This follows from Theorem 3.3 when T is CA. However, such an infinite set does not help us to answer Leibniz's Question.

3.2 The theory Q

We noted at the end of Subsection 2.2 that to be able to derive theorems of number theory, that is, sentences true in the standard interpretation, we need to introduce some axioms which express properties of this interpretation. The only sentences derivable in our formal system from no assumptions are those which are true in all interpretations; to be able to derive other sentences, we need to add some axioms which we can use as assumptions.

We thus seek a set S of sentences which we can use as axioms for number theory, and we then consider the theory $\mathrm{Th}(S)$ consisting of all the sentences that we can derive using the sentences in S as assumptions.

When it comes to choosing the axioms which make up S there are a number of criteria that we aim to satisfy.

1 All the axioms in S should be true in the standard interpretation $\mathscr{N}$.

2 There should be an algorithm to determine whether a given sentence is in S.

3 We should be able to derive as many theorems as possible from the axioms.

4 The set of axioms should be as simple as possible.

The first two of these criteria are precise requirements for our set S of axioms. By Theorem 3.2, criterion 1 will ensure that all the theorems derived from the axioms are true in the standard interpretation, so that (by the remark following Definition 3.4) the theory $\mathrm{Th}(S)$ is consistent. Since we require number theory to be consistent, criterion 1 is clearly an essential requirement for any set of sentences which could be considered as a set of axioms for it. Criterion 2 will ensure that there is an algorithm for deciding whether a purported proof is really a proof in our formal system of a formula ψ from assumptions which are all axioms in S. Thus criterion 2 is another essential requirement.

We combine an algorithm which checks whether the proof follows the rules and identifies the assumptions on which ψ depends with the algorithm which checks whether these assumptions are sentences in S.

Criteria 3 and 4 are not quite so precise and they pull us in opposite directions: criterion 3 suggests that a large set of axioms might be sensible whereas criterion 4 suggests that we should look for a small set. In fact, the ideal way to achieve criterion 3 would be to find a set of axioms which meets criteria 1 and 2, and which enables us to derive as theorems *all* sentences true in the standard interpretation. In *Unit 8*, we shall prove that this is not possible: it forms the content of Gödel's First Incompleteness Theorem. So it makes sense to settle for a simple set of axioms, thus satisfying criterion 4, which meets criteria 1 and 2 and also enables us to derive as many theorems as possible, thus satisfying criterion 3.

We can satisfy criterion 3 by taking our set of axioms to be Complete Arithmetic, but, as we shall see in *Unit 8*, this set does not satisfy criterion 2.

However, at this point we need to remember that our ultimate goal is to prove Gödel's Incompleteness Theorems and to answer Leibniz's and Hilbert's Questions. It turns out that we are able to make progress towards our goal by considering a set of axioms which certainly meets criteria 1, 2 and 4, but is so far from meeting criterion 3 that it is very easy to find sentences which are true in the standard interpretation but which cannot be derived from the axioms. With our ultimate goal in mind, therefore, we consider the theory Q which can be derived from the following seven axioms.

We shall explain why it is fruitful to consider such a weak set of axioms in *Unit 7*.

Definition 3.6 Axioms of the theory Q

$Q1$ $\forall x \forall y\,(x' = y' \rightarrow x = y)$

$Q2$ $\forall x\, \neg\, \mathbf{0} = x'$

$Q3$ $\forall x\,(\neg\, x = \mathbf{0} \rightarrow \exists y\, x = y')$

$Q4$ $\forall x\,(x + \mathbf{0}) = x$

$Q5$ $\forall x \forall y\,(x + y') = (x + y)'$

$Q6$ $\forall x\,(x \cdot \mathbf{0}) = \mathbf{0}$

$Q7$ $\forall x \forall y\,(x \cdot y') = ((x \cdot y) + x)$

We now ask you to check that this set of axioms satisfies criterion 1.

Problem 3.4

Write down the properties of the standard interpretation $\mathcal{N}$ expressed by the axioms of Q, and hence satisfy yourself that each of these axioms is true in this interpretation.

It follows from Theorem 3.2 and Problem 3.4 that the standard interpretation $\mathcal{N}$ is an interpretation of the theory Q. This is, of course, as we intended. The theory Q is thus consistent.

In fact the interpretations of the formal language given in Examples 3.5 and 3.6 of *Unit 4* are non-standard interpretations of Q. That is, all seven of the axioms of Q are true in both of these interpretations.

Problem 3.5

Check the examples, problems and additional exercises of Section 3 of *Unit 4* to see which of the axioms of Q have been shown to be true in (a) the interpretation $\mathcal{N}^*$ in Example 3.5 of *Unit 4* and (b) the interpretation $\mathcal{N}^{**}$ in Example 3.6 of *Unit 4*.

The remaining axioms will be shown to be true in Example 3.2.

Example 3.2

We show that all the axioms of Q are true in the interpretations $\mathcal{N}^*$ and $\mathcal{N}^{**}$ of Examples 3.5 and 3.6 of *Unit 4*.

It has already been noted in the solution to Problem 3.5 that $Q3$, $Q5$ and $Q7$ are true in both of these interpretations.

We now turn our attention to the other four axioms.

Axiom $Q1$ expresses the fact that the successor function is a one–one function. It is evident from the tables for the successor function in Examples 3.5 and 3.6 of *Unit 4* that this axiom is true in both interpretations.

Axiom $Q2$ expresses that there is no element in the domain of the interpretation whose successor is 0. Again, using the successor tables in Examples 3.5 and 3.6 of *Unit 4*, it is easily checked that this is true in both interpretations.

The truth of axiom $Q4$ in both interpretations can be seen from the first columns of the addition tables in Examples 3.5 and 3.6 of *Unit 4*. Likewise the first columns of the multiplication tables in these examples show that axiom $Q6$ is true in both cases.

Thus all of the seven axioms of Q are true in these two interpretations. Hence any sentence which is derivable from these axioms is true in both interpretations. ♦

Now let us consider criterion 2. As the given set of axioms of Q is finite, there is a very simple algorithm for determining whether a given formula ϕ is an axiom. Just compare ϕ with each of the seven axioms of Q in turn. If ϕ is identical with one of the sentences $Q1$ to $Q7$, then it is an axiom; otherwise, it is not. Thus criterion 2 is satisfied.

We trust that you will agree that the set of axioms of Q is very simple, so that criterion 4 is satisfied.

Now we show that criterion 3 is far from being met.

It follows from Example 3.2 and the Correctness Theorem that any sentence which is not true in the interpretation $\mathscr{N}^*$ cannot be derived from the axioms of Q. That is, such a sentence cannot be a theorem of Q. Likewise, a sentence which is not true in the interpretation $\mathscr{N}^{**}$ also cannot be a theorem of Q. Thus, from Section 3 of *Unit 4*, it follows that none of the following sentences, each of which is true in the standard interpretation $\mathscr{N}$, is a theorem of Q.

$$\forall x \forall y \, (x + y) = (y + x)$$
$$\forall x \neg x' = x$$
$$\forall x \, (\mathbf{0} \cdot x) = \mathbf{0}$$
$$\forall x \forall y \, (x \cdot y) = (y \cdot x)$$
$$\forall x \, (\mathbf{0} + x) = x$$

The relevant parts of Section 3 of *Unit 4* are, respectively, Example 3.7(a), Problem 3.5(a) and (c), and Additional Exercise 5(a) and (c).

Problem 3.6

Show that none of the following sentences, each of which is true in the standard interpretation $\mathscr{N}$, is a theorem of Q.

(a) $\forall x \forall y \forall z \, (x + (y + z)) = ((x + y) + z)$

(b) $\forall x \forall y \forall z \, (x \cdot (y \cdot z)) = ((x \cdot y) \cdot z)$

(c) $\forall x \forall y \forall z \, (x \cdot (y + z)) = ((x \cdot y) + (x \cdot z))$

It is clear from these examples that there are many sentences which are true under the standard interpretation $\mathscr{N}$ but which cannot be derived from the axioms of Q. In this sense Q is a very weak theory. None the less, it will emerge from the next unit, where we indicate which theorems can be derived in Q, that it is a very useful theory. As an appetizer, we show that a sentence which in the standard interpretation states that 'one plus one makes two' can be derived from the axioms of Q.

Example 3.3

We show that

$$\vdash_Q (\mathbf{0}' + \mathbf{0}') = \mathbf{0}''$$

1	(1)	$\forall x \, (x + \mathbf{0}) = x$	Ass
2	(2)	$\forall x \forall y \, (x + y') = (x + y)'$	Ass
2	(3)	$\forall y \, (\mathbf{0}' + y') = (\mathbf{0}' + y)'$	UE, 2
2	(4)	$(\mathbf{0}' + \mathbf{0}') = (\mathbf{0}' + \mathbf{0})'$	UE, 3
1	(5)	$(\mathbf{0}' + \mathbf{0}) = \mathbf{0}'$	UE, 1
1, 2	(6)	$(\mathbf{0}' + \mathbf{0}') = \mathbf{0}''$	Sub, 5, 4

On line 6 we have obtained the formula $(\mathbf{0}' + \mathbf{0}') = \mathbf{0}''$ by substituting the term $\mathbf{0}'$ for the term $(\mathbf{0}' + \mathbf{0})$ in the formula on line 4. This is a legitimate use of the Substitution Rule because the formula $(\mathbf{0}' + \mathbf{0}) = \mathbf{0}'$, which permits such a substitution, occurs on line 5.

The assumptions in force on the final line are both axioms of Q. Hence we have shown that the formula $(\mathbf{0}' + \mathbf{0}') = \mathbf{0}''$ is a theorem of Q and so we can write $\vdash_Q (\mathbf{0}' + \mathbf{0}') = \mathbf{0}''$. ♦

SUMMARY

We began the unit with some practical advice about finding formal proofs. We drew attention to the usefulness of the technique of proof by contradiction. Then we gave a list of techniques for finding formal proofs based on considering both the logical form of any assumptions and the logical form of the desired conclusion. We supplemented these techniques by a table of some standard cases of tautological consequence.

Next we completed the description of our formal system by describing two rules of proof for handling the identity symbol, the Identity Introduction Rule (II) and the Substitution Rule (Sub). We drew attention to the fact that our formal system satisfies both the Correctness Theorem and the Adequacy Theorem, though we did not prove these theorems.

Finally we considered the matter of adding to our formal system some axioms to express properties possessed by our standard interpretation of number theory. After a general discussion of axiom systems and theories, we introduced a particular theory Q which has only seven axioms. We checked that all the axioms of Q are true in the standard interpretation and we used the non-standard interpretations introduced in *Unit 4* to show that there are many sentences of our formal language which are true in the standard interpretation but which cannot be derived from the axioms of Q. None the less we claimed that Q is a useful theory to study. We shall make good this claim in *Units 7* and *8*.

OBJECTIVES

We list those topics on which we may set assessment questions to test your understanding of this unit.

After working through this unit you should be able to:

(a) understand and use the technique of proof by contradiction;

(b) understand the meaning and use of the rules II and Sub;

(c) construct simple formal proofs using all nine rules of proof, making use of the techniques set out in Subsection 1.2 and the standard cases of tautological consequence set out in Subsection 1.3;

(d) understand and use the Correctness and Adequacy Theorems;

(e) understand the meanings of the terms theory, axiom, theorem, consistency, completeness and interpretation of a theory;

(f) understand the definitions of Complete Arithmetic and of the theory Q;

(g) use non-standard interpretations of Q to show that given sentences are not theorems of Q.

ADDITIONAL EXERCISES

Most of these exercises provide further practice, should you feel you need it, in handling the main ideas in the unit on which you are likely to be assessed.

There are a few harder problems, labelled as such in the margin. These are harder than any of the problems you are likely to encounter in the assessment and are included solely as challenges for the interested student.

Section 1

1 (a) What is wrong with the following attempt to show that, for all formulas ϕ, the formula $\neg\exists v\, \phi$ can be derived from the formula $\forall v\, \neg\phi$?

Warning! This is not a correct proof!

1	(1)	$\forall v\, \neg\phi$	Ass
1	(2)	$\neg\phi$	UE, 1
3	(3)	$\exists v\, \phi$	Ass
4	(4)	ϕ	Ass
1, 4	(5)	$(\phi \,\&\, \neg\phi)$	Taut, 2, 4
1, 3	(6)	$(\phi \,\&\, \neg\phi)$	EH, 5
1	(7)	$(\exists v\, \phi \rightarrow (\phi \,\&\, \neg\phi))$	CP, 6
1	(8)	$\neg\exists v\, \phi$	Taut, 7

(b) Give a correct schematic proof to show that, for all formulas ϕ,

Harder problem

$$\forall v\, \neg\phi \vdash \neg\exists v\, \phi$$

2 Show how the proofs in Examples 1.1 and 1.3 can be used to help show that, for all formulas ϕ,

Harder problem

$$\vdash (\exists v\, \neg\phi \leftrightarrow \neg\forall v\, \phi)$$

3 Give formal proofs to show the following.

(a) $\forall x\, (x + \mathbf{0}) = x \vdash \exists x\, (\mathbf{0} + x) = x$

(b) $\exists x\, x = x' \vdash \exists y\, y = y'$

(c) $\forall x\, (x + x) = (x \cdot x)' \vdash \forall y\, \exists z\, (y + y) = z'$

(d) $\vdash (\exists x\, \forall y\, (x + y') = y \rightarrow \forall z\, \exists x\, (x + z'') = z')$

4 Show each of the following, for all formulas ϕ, ψ and θ.

(a) $\exists v\, (\phi \leftrightarrow \psi), \forall v\, \neg\phi \vdash \exists v\, \neg\psi$

(b) $\forall v\, (\phi \vee \psi) \vdash (\exists v\, (\phi \rightarrow \theta) \rightarrow \exists v\, (\theta \vee \psi))$

(c) $\forall v\, (\phi \rightarrow \theta), \exists v\, (\phi \,\&\, \psi) \vdash \neg\forall v\, (\neg\theta \leftrightarrow \psi)$

5 Show that, for all formulas γ, δ, ϕ and ψ,

Harder problem

$$(\delta \leftrightarrow \exists y\, (\chi \,\&\, \psi)), \forall y\, (\chi \leftrightarrow \gamma) \vdash (\delta \leftrightarrow \exists y\, (\gamma \,\&\, \psi))$$

This result is used in *Unit 7*.

6 Show that, for all formulas ϕ, ψ and θ,

$$\neg\phi, \exists v\, (\psi \rightarrow \phi) \vdash (\forall v\, (\theta \vee \psi) \rightarrow \neg\forall v(\theta \rightarrow \phi))$$

provided that the variable v does not occur freely in ϕ. Indicate the steps of your proof which require that v does not occur freely in ϕ.

7 Show that, for all formulas ϕ,

Harder problem

$$\forall v\, \phi \vdash \neg\exists v\, \neg\phi$$

Section 2

1 The following is a correct schematic proof from which the assumption numbers have been omitted. Fill in the missing assumption numbers.

(1)	$\forall x \, \forall y \, x = y$	Ass
(2)	$\forall y \, x = y$	UE, 1
(3)	$x = x$	II
(4)	$x = y$	UE, 2
(5)	$y = x$	Sub, 3, 4
(6)	$x = x$	UE, 2

2 Give formal proofs to show the following.

(a) $\vdash \forall x \, \forall y \, (x = y \to (x \cdot x) = (y \cdot y))$

(b) $\vdash \forall x \, \forall y \, \forall z \, (x = y \to (z \cdot x) = (z \cdot y))$

3 Give examples of formulas ϕ and ψ such that ψ is a logical consequence of ϕ but not a tautological consequence of ϕ.

4 Show that the following statements are not true.

(a) $\vdash \forall x \, (\neg x = \mathbf{0} \to \exists y \, x = y')$
Hint: Use one of the facts established in the examples and problems in Section 3 of *Unit 4*.

(b) $\vdash \forall x \, \forall y \, (x' = y' \to x = y)$

5 Show that it is not the case that, for all formulas ϕ,

$$\vdash (\exists v \, \phi \to \forall v \, \phi)$$

Section 3

1 Let T be a complete and consistent theory. Show that if Φ and Ψ are sentences such that $\vdash_T (\Phi \vee \Psi)$, then either $\vdash_T \Phi$ or $\vdash_T \Psi$.

2 (a) Suppose that $\Phi_1, \Phi_2, \ldots, \Phi_k$ are sentences and that Ψ is a sentence such that $\Phi_1, \Phi_2, \ldots, \Phi_k \vdash \Psi$. Show that

$$\vdash (\Phi_1 \to (\Phi_2 \to (\cdots \to (\Phi_k \to \Psi) \cdots)))$$

(b) Theorem 3.1 states that, for any theory T and all sentences Ψ, if Ψ is derivable from T then $\Psi \in T$. Prove this theorem directly within the formal system without appeal to the Correctness and Adequacy Theorems. *Hint:* As Ψ is derivable from T then there are sentences $\Phi_1, \Phi_2, \ldots, \Phi_k$ in T such that $\Phi_1, \Phi_2, \ldots, \Phi_k \vdash \Psi$. If T has axioms in the set S, there are derivations of each of $\Phi_1, \Phi_2, \ldots, \Phi_k$ with assumptions in the set S. Combine all these derivations together in a suitable way.

3 Show that the interpretation of the formal language given in Example 3.4 of Section 3 of *Unit 4* is not an interpretation of the theory Q. Deduce that the axiom $Q3$ is not a logical consequence of the other axioms of Q.

4 Show that none of the following sentences is a theorem of Q.

(a) $\exists y \, \forall x \, (y + x') = x''$

(b) $\forall x \, \forall y \, (x'' + y) = (y' + x')$

(c) $\exists y \, \forall x \, (y \cdot x) = y$

SOLUTIONS TO THE PROBLEMS

Solution 1.1

(a)

1	(1)	$\exists v\, \phi$	Ass
2	(2)	ϕ	Ass
3	(3)	$\forall v\, \neg\phi$	Ass
3	(4)	$\neg\phi$	UE, 3
2, 3	(5)	$(\phi \,\&\, \neg\phi)$	Taut, 2, 4
2	(6)	$(\forall v\, \neg\phi \rightarrow (\phi \,\&\, \neg\phi))$	CP, 5
2	(7)	$\neg\forall v\, \neg\phi$	Taut, 6
1	(8)	$\neg\forall v\, \neg\phi$	EH, 7

The first two lines are standard when one of the assumptions is an existential formula. We then assume $\forall v\, \neg\phi$ hoping to derive a contradiction. We use the UE Rule to drop the universal quantifier from this formula, derive a contradiction and then use the proof by contradiction method. The use of the EH Rule on line 8 is legitimate as the variable v cannot occur freely in the formula $\neg\forall v\, \neg\phi$.

This sort of commentary on the proof would not normally be required in the answer to an assessment question. A correct formal proof would normally suffice on its own!

(b)

1	(1)	$\forall v\, \phi$	Ass
2	(2)	$\exists v\, \neg\psi$	Ass
3	(3)	$\neg\psi$	Ass
1	(4)	ϕ	UE, 1
5	(5)	$\forall v\, (\phi \rightarrow \psi)$	Ass
5	(6)	$(\phi \rightarrow \psi)$	UE, 5
1, 5	(7)	ψ	Taut, 4, 6
1, 3, 5	(8)	$(\psi \,\&\, \neg\psi)$	Taut, 3, 7
1, 3	(9)	$(\forall v\, (\phi \rightarrow \psi) \rightarrow (\psi \,\&\, \neg\psi))$	CP, 8
1, 3	(10)	$\neg\forall v\, (\phi \rightarrow \psi)$	Taut, 9
1, 2	(11)	$\neg\forall v\, (\phi \rightarrow \psi)$	EH, 10

The first four lines are standard. We introduce the given assumptions on lines 1 and 2. As one of the assumptions is an existential formula, we also assume the formula on line 3. As another is a universal formula, we use the UE Rule to drop the universal quantifier. We then assume the formula $\forall v\, (\phi \rightarrow \psi)$, use the UE Rule to drop the universal quantifier from this extra assumption and try to derive a contradiction. After deriving a contradiction, we use the proof by contradiction method. The use of the EH Rule on line 11 is legitimate as the variable v cannot occur freely in either the formula $\neg\forall v\, (\phi \rightarrow \psi)$ being derived or the other assumption $\forall v\, \phi$ on line 10.

Solution 1.2

(a)

1	(1)	$\forall v\, (\neg\phi \rightarrow \psi)$	Ass
2	(2)	$\neg\exists v\, \psi$	Ass
1	(3)	$(\neg\phi \rightarrow \psi)$	UE, 1
4	(4)	$\neg\phi$	Ass
1, 4	(5)	ψ	Taut, 3, 4
1, 4	(6)	$\exists v\, \psi$	EI, 5
1, 2, 4	(7)	$(\exists v\, \psi \,\&\, \neg\exists v\, \psi)$	Taut, 2, 6
1, 2	(8)	$(\neg\phi \rightarrow (\exists v\, \psi \,\&\, \neg\exists v\, \psi))$	CP, 7
1, 2	(9)	ϕ	Taut, 8
1, 2	(10)	$\forall v\, \phi$	UI, 9

The first three lines of the above schematic proof are standard. First we introduce the two given assumptions and then use the UE Rule to remove the universal quantifier from the first assumption. Then we turn our attention to the formula $\forall v\, \phi$ that we are ultimately aiming to derive. Our strategy is first to derive ϕ and then use the UI Rule. Since there is no other obvious way to make progress towards our aim of deriving ϕ, we introduce $\neg\phi$ as an assumption with a view to deriving a contradiction. Keeping in mind that we have the formula $\neg\exists v\,\psi$ on line 2, we know that, to derive a contradiction, it would be enough to derive $\exists v\,\psi$. We then use the proof by contradiction method.

(b)

1	(1)	ϕ	Ass
2	(2)	$\exists v\,(\psi \to \neg\phi)$	Ass
3	(3)	$(\psi \to \neg\phi)$	Ass
4	(4)	$\forall v\,\psi$	Ass
4	(5)	ψ	UE, 4
3, 4	(6)	$\neg\phi$	Taut, 3, 5
1, 3, 4	(7)	$(\phi \,\&\, \neg\phi)$	Taut, 1, 6
1, 2, 4	(8)	$(\phi \,\&\, \neg\phi)$	EH, 7
1, 2	(9)	$(\forall v\,\psi \to (\phi \,\&\, \neg\phi))$	CP, 8
1, 2	(10)	$\neg\forall v\,\psi$	Taut, 9

As $\neg\forall v\,\psi$ is a negation, our strategy is to add the assumption $\forall v\,\psi$ to the assumptions ϕ, $\exists v\,(\psi \to \neg\phi)$ and try to derive a contradiction. As the assumption $\exists v\,(\psi \to \neg\phi)$ is an existential formula, preparatory to a use of the EH Rule we also assume the formula $(\psi \to \neg\phi)$. We use the UE Rule on line 5 to drop the universal quantifier from one of the assumptions and then derive a contradiction on line 7. Because of the condition that the variable v does not occur freely in ϕ, we have taken care to derive the contradiction $(\phi \,\&\, \neg\phi)$, rather than say $(\psi \,\&\, \neg\psi)$, so that we can then use the EH Rule on line 8 and finish off the proof by contradiction.

Solution 1.3

Here we adopt the standard strategy for proving disjunctions. To derive $(\exists v\,\phi \vee \forall v\,\neg\phi)$ we use the fact that this formula is a tautological consequence of the implication $(\neg\exists v\,\phi \to \forall v\,\neg\phi)$. We can derive this implication by first deriving $\forall v\,\neg\phi$ from the assumption $\neg\exists v\,\phi$ and then using the CP Rule. The derivation of $\forall v\,\neg\phi$ from $\neg\exists v\,\phi$ is very similar to our derivation of $\forall v\,\phi$ from $\neg\exists v\,\neg\phi$ in Example 1.3. We thus arrive at the following schematic proof.

1	(1)	$\neg\exists v\,\phi$	Ass
2	(2)	ϕ	Ass
2	(3)	$\exists v\,\phi$	EI, 2
1, 2	(4)	$(\exists v\,\phi \,\&\, \neg\exists v\,\phi)$	Taut, 1, 3
1	(5)	$(\phi \to (\exists v\,\phi \,\&\, \neg\exists v\,\phi))$	CP, 4
1	(6)	$\neg\phi$	Taut, 5
1	(7)	$\forall v\,\neg\phi$	UI, 6
	(8)	$(\neg\exists v\,\phi \to \forall v\,\neg\phi)$	CP, 7
	(9)	$(\exists v\,\phi \vee \forall v\,\neg\phi)$	Taut, 8

Solution 1.4

We follow the standard strategy for proving bi-implications.

You can use truth tables to check that the use of the Tautology Rule on lines 2 and 5 is correct.

1	(1)	$\neg(\phi \,\&\, \psi)$	Ass
1	(2)	$(\neg\phi \vee \neg\psi)$	Taut, 1
	(3)	$(\neg(\phi \,\&\, \psi) \to (\neg\phi \vee \neg\psi))$	CP, 2
4	(4)	$(\neg\phi \vee \neg\psi)$	Ass
4	(5)	$\neg(\phi \,\&\, \psi)$	Taut, 4
	(6)	$((\neg\phi \vee \neg\psi) \to \neg(\phi \,\&\, \psi))$	CP, 5
	(7)	$(\neg(\phi \,\&\, \psi) \leftrightarrow (\neg\phi \vee \neg\psi))$	Taut, 3, 6

Solution 1.5

1	(1)	$\exists v\,(\phi \rightarrow \psi)$	Ass
2	(2)	$\forall v\,\phi$	Ass
3	(3)	$(\phi \rightarrow \psi)$	Ass
2	(4)	ϕ	UE, 2
2, 3	(5)	ψ	Taut, 3, 4
2, 3	(6)	$\exists v\,\psi$	EI, 5
1, 2	(7)	$\exists v\,\psi$	EH, 6

This proof starts with a straightforward use of techniques (T1), (T2) and (T3) to give lines 1 to 4. As we want to derive $\exists v\,\psi$, we follow (T10) and first try to derive ψ, which we can by using the Tautology Rule.

Solution 1.6

1	(1)	$\forall v\,\phi$	Ass
2	(2)	$\exists v\,(\psi \rightarrow \neg\phi)$	Ass
1	(3)	ϕ	UE, 1
4	(4)	$(\psi \rightarrow \neg\phi)$	Ass
5	(5)	$\forall v\,\psi$	Ass
5	(6)	ψ	UE, 5
1, 4	(7)	$\neg\psi$	Taut, 3, 4
1, 4, 5	(8)	$(\psi \,\&\, \neg\psi)$	Taut, 6, 7
1, 4	(9)	$(\forall v\,\psi \rightarrow (\psi \,\&\, \neg\psi))$	CP, 8
1, 4	(10)	$\neg\forall v\,\psi$	Taut, 9
1, 2	(11)	$\neg\forall v\,\psi$	EH, 10

One alternative proof would have lines 1 to 6 as here, but on line 7 would apply the Tautology Rule to lines 4 and 6 to obtain $\neg\phi$, depending on assumptions 4 and 5. The contradiction on line 8 would then be $(\phi \,\&\, \neg\phi)$ and on line 9 we would then have $(\forall v\,\psi \rightarrow (\phi \,\&\, \neg\phi))$.

We obtained this schematic proof by following standard strategies. Following (T1), the assumptions we are given are introduced on the first two lines. Following (T2) and (T3), the UE Rule is applied to the first assumption, and we introduce $(\psi \rightarrow \neg\phi)$ as a new assumption on line 4 with a view to a later use of the EH Rule. Now we turn our attention to the desired conclusion, $\neg\forall v\,\psi$. Since this begins with a negation symbol, we follow (T4) by introducing the assumption $\forall v\,\psi$ and aiming to derive a contradiction. The UE Rule applied to $\forall v\,\psi$ gives us the formula ψ. Thus to get a contradiction all we now need do is derive $\neg\psi$. It is not difficult to see that this is a tautological consequence of the formulas ϕ and $(\psi \rightarrow \neg\phi)$. We are thus able to complete the proof in the intended way.

Solution 1.7

1	(1)	$\forall v\,(\phi \rightarrow \theta)$	Ass
2	(2)	$\exists v\,(\psi \vee \neg\phi)$	Ass
1	(3)	$(\phi \rightarrow \theta)$	UE, 1
4	(4)	$(\psi \vee \neg\phi)$	Ass
5	(5)	$\forall v\,\phi$	Ass
5	(6)	ϕ	UE, 5
7	(7)	$\forall v\,\neg(\psi \,\&\, \theta)$	Ass
7	(8)	$\neg(\psi \,\&\, \theta)$	UE, 7
4, 5	(9)	ψ	Taut, 4, 6
1, 5	(10)	θ	Taut, 3, 6
1, 4, 5	(11)	$(\psi \,\&\, \theta)$	Taut, 9, 10
1, 4, 5, 7	(12)	$((\psi \,\&\, \theta) \,\&\, \neg(\psi \,\&\, \theta))$	Taut, 8, 11
1, 4, 5	(13)	$(\forall v\,\neg(\psi \,\&\, \theta) \rightarrow ((\psi \,\&\, \theta) \,\&\, \neg(\psi \,\&\, \theta)))$	CP, 12
1, 4, 5	(14)	$\neg\forall v\,\neg(\psi \,\&\, \theta)$	Taut, 13
1, 2, 5	(15)	$\neg\forall v\,\neg(\psi \,\&\, \theta)$	EH, 14
1, 2	(16)	$(\forall v\,\phi \rightarrow \neg\forall v\,\neg(\psi \,\&\, \theta))$	CP, 15

This is quite a long schematic proof, but it can be seen that we have obtained it by following standard strategies. Following (T1), the two given assumption formulas are introduced and handled in the standard way on the first four lines, following (T2) and (T3). Then we turn our attention to the desired conclusion. Since this is the implication $(\forall v\, \phi \rightarrow \neg\forall v\, \neg(\psi \,\&\, \theta))$, we follow (T7) and introduce the antecedent, $\forall v\, \phi$, as an additional assumption on line 5 and aim to derive the consequent $\neg\forall v\, \neg(\psi \,\&\, \theta)$. Since this is a negation, we follow (T4) by introducing $\forall v\, \neg(\psi \,\&\, \theta)$ as a new assumption with the aim of deriving a contradiction. We achieve a contradiction on line 12 and then complete the proof in accordance with the strategies that we have already mentioned.

Solution 1.8

We need to show that, whenever all the formulas ϕ, $(\neg\phi \vee \psi)$ and $\neg\theta$ are true, then the formula $\neg(\psi \rightarrow \theta)$ is also true. So we need only consider the lines of the truth table where ϕ and $\neg\theta$ are both true, that is where ϕ is true and θ is false. So we need only consider two lines of the full truth table.

$(((\phi$	$\&$	$(\neg$	ϕ	$\vee$	$\psi))$	$\&$	$\neg$	$\theta)$	$\rightarrow$	$\neg$	$(\psi$	$\rightarrow$	$\theta))$
1	1	0	1	1	1	1	1	0	1	1	1	0	0
1	0	0	1	0	0	0	1	0	1	0	0	1	0
①	④	②	①	③	①	⑤	②	①	⑥	③	①	②	①

We see from column ⑥ that the formula is a tautology. It follows that $\neg(\psi \rightarrow \theta)$ is a tautological consequence of the formulas ϕ, $(\neg\phi \vee \psi)$ and $\neg\theta$.

Solution 2.1

(a)

1	(1)	$x = y$	Ass
	(2)	$(z + x) = (z + x)$	II
1	(3)	$(z + x) = (z + y)$	Sub, 1, 2
	(4)	$(x = y \rightarrow (z + x) = (z + y))$	CP, 3
	(5)	$\forall z\, (x = y \rightarrow (z + x) = (z + y))$	UI, 4
	(6)	$\forall y\, \forall z\, (x = y \rightarrow (z + x) = (z + y))$	UI, 5
	(7)	$\forall x\, \forall y\, \forall z\, (x = y \rightarrow (z + x) = (z + y))$	UI, 6

The formula on line 3 has been obtained from that on line 2 by substituting y for the second occurrence of x. Since we have the formula $x = y$ on line 1, this is a legitimate use of the Substitution Rule.

(b)

1	(1)	$x = y$	Ass
2	(2)	$y = z$	Ass
1, 2	(3)	$x = z$	Sub, 1, 2
1	(4)	$(y = z \rightarrow x = z)$	CP, 3
	(5)	$(x = y \rightarrow (y = z \rightarrow x = z))$	CP, 4
	(6)	$\forall z\, (x = y \rightarrow (y = z \rightarrow x = z))$	UI, 5
	(7)	$\forall y\, \forall z\, (x = y \rightarrow (y = z \rightarrow x = z))$	UI, 6
	(8)	$\forall x\, \forall y\, \forall z\, (x = y \rightarrow (y = z \rightarrow x = z))$	UI, 7

The formula on line 3 has been obtained from the formula on line 1 by substituting z for y. This substitution is a legitimate use of the Substitution Rule because we have the formula $y = z$ on line 2.

Note that it is not a legitimate use of the Substitution Rule to substitute x for y in the formula on line 2, as the formula $x = y$ on line 1 does not permit substitution of x for y. It only permits substitution of y for x.

Solution 2.2

(a)

1	(1)	$y = x$	Ass
	(2)	$y' = y'$	II
1	(3)	$x' = y'$	Sub, 1, 2

(b)

1	(1)	$x = y$	Ass
	(2)	$((x \cdot x) \cdot x) = ((x \cdot x) \cdot x)$	II
1	(3)	$((y \cdot x) \cdot y) = ((x \cdot y) \cdot x)$	Sub, 1, 2

Solution 2.3

We need to find an example of a formula ϕ, and an interpretation, for which the formula $(\forall x\, \exists y\, \phi \rightarrow \exists y\, \forall x\, \phi)$ is not true. It will then follow from the Correctness Theorem that it is not the case that $\vdash (\forall x\, \exists y\, \phi \rightarrow \exists y\, \forall x\, \phi)$.

There are lots of possible choices for the formula ϕ. We take ϕ to be the formula $y = x'$. In the standard interpretation, the formula $\forall x\, \exists y\, y = x'$ is interpreted as 'for every natural number x, there is a natural number y which is the successor of x' and this is true. On the other hand the formula $\exists y\, \forall x\, y = x'$ is interpreted as 'there is a natural number y which is the successor of every natural number x', which is false. So with this choice of ϕ, the formula $(\forall x\, \exists y\, \phi \rightarrow \exists y\, \forall x\, \phi)$ is false in the standard interpretation.

Solution 3.1

Plainly every sentence Ψ in S is derivable from S using the one-line proof

1 (1) Ψ Ass

So each such Ψ is in $\mathrm{Th}(S)$.

Equivalently, if $\Psi \in S$, then every interpretation which makes S true also makes Ψ true, so Ψ is a logical consequence of S and hence $\Psi \in \mathrm{Th}(S)$.

Solution 3.2

Suppose that T is a theory which is not consistent. Then there is some sentence Φ of the formal language such that both $\vdash_T \Phi$ and $\vdash_T \neg\Phi$. Now let Ψ be any sentence of the formal language. The formula $((\Phi \,\&\, \neg\Phi) \rightarrow \Psi)$ is a tautology. Hence Ψ can be derived from the formulas Φ and $\neg\Phi$ using the Tautology Rule. Therefore, as T is a theory and so is closed under derivation (Theorem 3.1), it follows that $\vdash_T \Psi$.

Solution 3.3

Suppose that T is a theory which is both complete and consistent. Let Φ be any sentence of the formal language. Since T is complete, either $\vdash_T \Phi$ or $\vdash_T \neg\Phi$. Since T is consistent, we cannot have both $\vdash_T \Phi$ and $\vdash_T \neg\Phi$. Hence precisely one of $\vdash_T \Phi$ and $\vdash_T \neg\Phi$ holds.

Solution 3.4

We consider each of the seven axioms of Q in turn and state what they express in the standard interpretation $\mathscr{N}$. It should be evident in each case that the axiom expresses a true statement about $\mathscr{N}$.

$Q1$ expresses that the function succ, the successor function, is one–one.

Recall that a function $f : A \longrightarrow B$ is one–one if, for all $a_1, a_2 \in A$,

$$f(a_1) = f(a_2) \Rightarrow a_1 = a_2.$$

$Q2$ expresses the fact that 0 is not the successor of any natural number.

$Q3$ expresses the fact that every non-zero natural number is the successor of some natural number.

$Q4$ expresses the fact that, for every natural number n,

$$n + 0 = n.$$

$Q5$ expresses the fact that, for all natural numbers n and m,

$$n + \mathrm{succ}(m) = \mathrm{succ}(n + m).$$

Note that the equations corresponding to axioms $Q4$ and $Q5$ give the definition of addition by primitive recursion, as given in *Unit 2*.

$Q6$ expresses the fact that, for every natural number n,

$$n \times 0 = 0.$$

$Q7$ expresses the fact that, for all natural numbers n and m,

$$n \times \mathrm{succ}(m) = (n \times m) + n.$$

Note that the equations corresponding to axioms $Q6$ and $Q7$ give the definition of multiplication by primitive recursion, as given in *Unit 2*.

Solution 3.5

In Example 3.7(b) of *Unit 4* it is shown that axiom $Q3$ is true both in $\mathscr{N}^*$ and $\mathscr{N}^{**}$. It follows from Solution 3.5(b) of *Unit 4* that axiom $Q5$ is true in both these interpretations, and from the solution to part (b) of Additional Exercise 5 for Section 3 *of Unit 4* that Axiom $Q7$ is true in both these interpretations.

Solution 3.6

We show that in each case the sentence is not true in the interpretation $\mathscr{N}^{**}$ of Example 3.6 of *Unit 4*. As all the axioms of Q are true in this interpretation, it will follow from the Correctness Theorem that the sentence is not a theorem of Q.

Since each sentence expresses a universal property, we can show it is false by giving a single counter-example in each case. As the following calculations show, we can obtain a counter-example for each sentence by interpreting x, y and z as the element α of the interpretation $\mathscr{N}^{**}$.

There are other interpretations of x, y and z which will give counter-examples, for instance interpreting them all by β.

(a) $\alpha + (\alpha + \alpha) = \alpha + \beta = \alpha$, whereas $(\alpha + \alpha) + \alpha = \beta + \alpha = \beta$.
Hence $\alpha + (\alpha + \alpha) \neq (\alpha + \alpha) + \alpha$. So the sentence
$\forall x \forall y \forall z\, (x + (y + z)) = ((x + y) + z)$ is not true in $\mathscr{N}^{**}$.

(b) $\alpha \cdot (\alpha \cdot \alpha) = \alpha \cdot \beta = \beta$, whereas $(\alpha \cdot \alpha) \cdot \alpha = \beta \cdot \alpha = \alpha$.
Hence $\alpha \cdot (\alpha \cdot \alpha) \neq (\alpha \cdot \alpha) \cdot \alpha$. So the sentence
$\forall x \forall y \forall z\, (x \cdot (y \cdot z)) = ((x \cdot y) \cdot z)$ is not true in $\mathscr{N}^{**}$.

(c) $\alpha \cdot (\alpha + \alpha) = \alpha \cdot \beta = \beta$, whereas $(\alpha \cdot \alpha) + (\alpha \cdot \alpha) = \beta + \beta = \alpha$.
Hence $\alpha \cdot (\alpha + \alpha) \neq (\alpha \cdot \alpha) + (\alpha \cdot \alpha)$. So the sentence
$\forall x \forall y \forall z\, (x \cdot (y + z)) = ((x \cdot y) + (x \cdot z))$ is not true in $\mathscr{N}^{**}$.

SOLUTIONS TO ADDITIONAL EXERCISES

Section 1

1 (a) The use of the EH Rule on line 6 is not correct as, in general, the variable v may occur freely in the formula $(\phi \,\&\, \neg\phi)$ on line 5.

(b) We can overcome this difficulty by using the contradiction on line 5 to derive a contradiction in which the variable v does not occur freely. We thus arrive at the following correct proof.

1	(1)	$\forall v\, \neg\phi$	Ass
1	(2)	$\neg\phi$	UE, 1
3	(3)	$\exists v\, \phi$	Ass
4	(4)	ϕ	Ass
1, 4	(5)	$(\phi \,\&\, \neg\phi)$	Taut, 2, 4
4	(6)	$(\forall v\, \neg\phi \rightarrow (\phi \,\&\, \neg\phi))$	CP, 5
4	(7)	$\neg\forall v\, \neg\phi$	Taut, 6
1, 4	(8)	$(\forall v\, \neg\phi \,\&\, \neg\forall v\, \neg\phi)$	Taut, 1, 7
1, 3	(9)	$(\forall v\, \neg\phi \,\&\, \neg\forall v\, \neg\phi)$	EH, 8
1	(10)	$(\exists v\, \phi \rightarrow (\forall v\, \neg\phi \,\&\, \neg\forall v\, \neg\phi))$	CP, 9
1	(11)	$\neg\exists v\, \phi$	Taut, 10

2 Technique (T8) tells us that to derive the bi-implication $(\exists v \neg\phi \leftrightarrow \neg\forall v\, \phi)$ we should aim first to derive the two implications $(\exists v \neg\phi \rightarrow \neg\forall v\, \phi)$ and $(\neg\forall v\, \phi \rightarrow \exists v \neg\phi)$. The first of these implications can be obtained from the formal proof of Example 1.1 by one use of the CP Rule. The second implication can be derived by noting first that it is a tautological consequence of $(\neg\exists v \neg\phi \rightarrow \forall v\, \phi)$ and then that this can be derived from the formal proof of Example 1.3 by one use of the CP Rule. We thus obtain the following formal proof.

1	(1)	$\exists v \neg\phi$	Ass
2	(2)	$\neg\phi$	Ass
3	(3)	$\forall v\, \phi$	Ass
3	(4)	ϕ	UE, 3
2, 3	(5)	$(\phi \,\&\, \neg\phi)$	Taut, 2, 4
2	(6)	$(\forall v\, \phi \rightarrow (\phi \,\&\, \neg\phi))$	CP, 5
2	(7)	$\neg\forall v\, \phi$	Taut, 6
1	(8)	$\neg\forall v\, \phi$	EH, 7
	(9)	$(\exists v \neg\phi \rightarrow \neg\forall v\, \phi)$	CP, 8
10	(10)	$\neg\exists v \neg\phi$	Ass
11	(11)	$\neg\phi$	Ass
11	(12)	$\exists v \neg\phi$	EI, 11
10, 11	(13)	$(\exists v \neg\phi \,\&\, \neg\exists v \neg\phi)$	Taut, 10, 12
10	(14)	$(\neg\phi \rightarrow (\exists v \neg\phi \,\&\, \neg\exists v \neg\phi))$	CP, 13
10	(15)	ϕ	Taut, 14
10	(16)	$\forall v\, \phi$	UI, 15
	(17)	$(\neg\exists v \neg\phi \rightarrow \forall v\, \phi)$	CP, 16
	(18)	$(\neg\forall v\, \phi \rightarrow \exists v \neg\phi)$	Taut, 17
	(19)	$(\exists v \neg\phi \leftrightarrow \neg\forall v\, \phi)$	Taut, 9, 18

3 (a) Technique (T10) suggests that we should look for a term τ such that we can derive the formula $\phi(\tau/x)$ where ϕ is the formula $(\mathbf{0} + x) = x$. We note that $\phi(\mathbf{0}/x)$ is the formula $(\mathbf{0} + \mathbf{0}) = \mathbf{0}$, which can easily be derived from the formula $\forall x\, (x + \mathbf{0}) = x$. So we obtain the following formal proof.

1	(1)	$\forall x\, (x + \mathbf{0}) = x$	Ass
1	(2)	$(\mathbf{0} + \mathbf{0}) = \mathbf{0}$	UE, 1
1	(3)	$\exists x\, (\mathbf{0} + x) = x$	EI, 2

(b)

1	(1)	$\exists x\, x = x'$	Ass
2	(2)	$x = x'$	Ass
2	(3)	$\exists y\, y = y'$	EI, 2
1	(4)	$\exists y\, y = y'$	EH, 3

The proof starts with the standard preparation for a use of the EH Rule and we then follow (T10).

(c)

1	(1)	$\forall x\, (x + x) = (x \cdot x)'$	Ass
1	(2)	$(y + y) = (y \cdot y)'$	UE, 1
1	(3)	$\exists z\, (y + y) = z'$	EI, 2
1	(4)	$\forall y\, \exists z\, (y + y) = z'$	UI, 3

(T9) suggests that we try to derive the formula $\exists z\, (y + y) = z'$. To derive this, (T10) suggests we first derive $(y + y) = \tau'$ for some term τ. This in turn is what leads us to eliminate the $\forall x$ in the assumption on line 1 by replacing the xs by ys.

(d)

1	(1)	$\exists x\,\forall y\,(x+y')=y$	Ass
2	(2)	$\forall y\,(x+y')=y$	Ass
2	(3)	$(x+z'')=z'$	UE, 2
2	(4)	$\exists x\,(x+z'')=z'$	EI, 3
2	(5)	$\forall z\,\exists x\,(x+z'')=z'$	UI, 4
1	(6)	$\forall z\,\exists x\,(x+z'')=z'$	EH, 5
	(7)	$(\exists x\,\forall y\,(x+y')=y \rightarrow \forall z\,\exists x\,(x+z'')=z')$	CP, 6

Note that there is a correct alternative proof based on using the EH Rule straight after line 4 to obtain $\exists x\,(x+z'')=z'$ on line 5 depending on assumption 1. Line 6 would then follow using the UI Rule.

Following (T7), we assume the formula $\exists x\,\forall y\,(x+y')=y$, derive $\forall z\,\exists x\,(x+z'')=z'$ and use the CP Rule. Lines 1 and 2 are then the standard preparation for a use of the EH Rule. We can anticipate following (T10) and then (T9) once we have derived the formula $(x+z'')=z'$, which we can do on line 3 with a straightforward use of the UE Rule.

4 (a)

1	(1)	$\exists v\,(\phi \leftrightarrow \psi)$	Ass
2	(2)	$(\phi \leftrightarrow \psi)$	Ass
3	(3)	$\forall v\,\neg\phi$	Ass
3	(4)	$\neg\phi$	UE, 3
2, 3	(5)	$\neg\psi$	Taut, 2, 4
2, 3	(6)	$\exists v\,\neg\psi$	EI, 5
1, 3	(7)	$\exists v\,\neg\psi$	EH, 6

As usual, we begin by introducing the given assumptions and, as one of them is an existential formula, we follow technique (T3). Following (T10), we aim to derive $\neg\psi$. A simple use of the UE Rule, following (T2), and use of the Tautology Rule does indeed give the formula $\neg\psi$. The desired conclusion follows by routine applications of the EI and EH Rules.

(b)

1	(1)	$\forall v\,(\phi \vee \psi)$	Ass
2	(2)	$\exists v\,(\phi \rightarrow \theta)$	Ass
3	(3)	$(\phi \rightarrow \theta)$	Ass
1	(4)	$(\phi \vee \psi)$	UE, 1
1, 3	(5)	$(\theta \vee \psi)$	Taut, 3, 4
1, 3	(6)	$\exists v\,(\theta \vee \psi)$	EI, 5
1, 2	(7)	$\exists v\,(\theta \vee \psi)$	EH, 6
1	(8)	$(\exists v\,(\phi \rightarrow \theta) \rightarrow \exists v\,(\theta \vee \psi))$	CP, 7

We follow (T7) and introduce the extra assumption $\exists v\,(\phi \rightarrow \theta)$. We then follow (T2) and (T3) for dealing with assumptions which are, respectively, universal and existential formulas. With (T10) in mind, we seek to derive $(\theta \vee \psi)$, which follows from the Tautology Rule.

(c)

1	(1)	$\forall v\,(\phi \to \theta)$	Ass
2	(2)	$\exists v\,(\phi \,\&\, \psi)$	Ass
3	(3)	$(\phi \,\&\, \psi)$	Ass
1	(4)	$(\phi \to \theta)$	UE, 1
5	(5)	$\forall v\,(\neg\theta \leftrightarrow \psi)$	Ass
5	(6)	$(\neg\theta \leftrightarrow \psi)$	UE, 5
1, 3	(7)	θ	Taut, 3, 4
3, 5	(8)	$\neg\theta$	Taut, 3, 6
1, 3, 5	(9)	$(\theta \,\&\, \neg\theta)$	Taut, 7, 8
1, 3	(10)	$(\forall v\,(\neg\theta \leftrightarrow \psi) \to (\theta \,\&\, \neg\theta))$	CP, 9
1, 3	(11)	$\neg\forall v\,(\neg\theta \leftrightarrow \psi)$	Taut, 10
1, 2	(12)	$\neg\forall v\,(\neg\theta \leftrightarrow \psi)$	EH, 11

The first four lines follow (T1), (T2) and (T3). As the desired conclusion starts with a negation, we follow (T4) in line 5 to set up a proof by contradiction, and then (T2). Use of the Tautology Rule then gives a contradiction, and the remainder of the derivation follows the method of proof by contradiction, with a final use of the EH Rule.

5

1	(1)	$(\delta \leftrightarrow \exists y\,(\chi \,\&\, \psi)$	Ass
2	(2)	$\forall y\,(\chi \leftrightarrow \gamma)$	Ass
2	(3)	$(\chi \leftrightarrow \gamma)$	UE, 2
4	(4)	$\exists y\,(\chi \,\&\, \psi)$	Ass
5	(5)	$(\chi \,\&\, \psi)$	Ass
2, 5	(6)	$(\gamma \,\&\, \psi)$	Taut, 3, 5
2, 5	(7)	$\exists y\,(\gamma \,\&\, \psi)$	EI, 6
2, 4	(8)	$\exists y\,(\gamma \,\&\, \psi)$	EH, 7
2	(9)	$(\exists y\,(\chi \,\&\, \psi) \to \exists y\,(\gamma \,\&\, \psi))$	CP, 8
1, 2	(10)	$(\delta \to \exists y\,(\gamma \,\&\, \psi))$	Taut, 1, 9
11	(11)	$\exists y\,(\gamma \,\&\, \psi)$	Ass
12	(12)	$(\gamma \,\&\, \psi)$	Ass
2, 12	(13)	$(\chi \,\&\, \psi)$	Taut, 3, 12
2, 12	(14)	$\exists y\,(\chi \,\&\, \psi)$	EI, 13
2, 11	(15)	$\exists y\,(\chi \,\&\, \psi)$	EH, 14
2	(16)	$(\exists y\,(\gamma \,\&\, \psi) \to \exists y\,(\chi \,\&\, \psi))$	CP, 15
1, 2	(17)	$(\exists y\,(\gamma \,\&\, \psi) \to \delta)$	Taut, 1, 16
1, 2	(18)	$(\delta \leftrightarrow \exists y\,(\gamma \,\&\, \psi))$	Taut, 10, 17

The first three lines follow (T1) and (T2). Technique (T8) then suggests that we try to derive $(\delta \to \exists y\,(\delta \,\&\, \psi))$ and $(\exists y\,(\gamma \,\&\, \psi) \to \delta)$ separately. To derive the former we follow (T7), though in light of the bi-implication on line 1 we introduce $\exists y\,(\chi \,\&\, \psi)$ as an assumption rather than δ. We then follow (T3), (T11) and (T10), before using the EH Rule and CP Rule to obtain lines 8 and 9, as suggested by our earlier use of (T3) and (T7). We then obtained the desired implication $\delta \to \exists y\,(\delta \,\&\, \psi)$ by use of the Tautology Rule again. We obtain the implication $\exists y\,(\gamma \,\&\, \psi) \to \delta$ in a similar fashion. Then we use the Tautology Rule, as suggested by our earlier use of (T8), to obtain the desired result.

6

1	(1)	$\neg\phi$	Ass
2	(2)	$\exists v\,(\psi \to \phi)$	Ass
3	(3)	$(\psi \to \phi)$	Ass
4	(4)	$\forall v\,(\theta \vee \psi)$	Ass
4	(5)	$(\theta \vee \psi)$	UE, 4
6	(6)	$\forall v\,(\theta \to \phi)$	Ass
6	(7)	$(\theta \to \phi)$	UE, 6
3, 4, 6	(8)	ϕ	Taut, 3, 5, 7
1, 3, 4, 6	(9)	$(\phi \mathbin{\&} \neg\phi)$	Taut, 1, 8
1, 3, 4	(10)	$(\forall v\,(\theta \to \phi) \to (\phi \mathbin{\&} \neg\phi))$	CP, 9
1, 3, 4	(11)	$\neg\forall v\,(\theta \to \phi)$	Taut, 10
1, 2, 4	(12)	$\neg\forall v\,(\theta \to \phi)$	EH, 11
1, 2	(13)	$(\forall v\,(\theta \vee \psi) \to \neg\forall v\,(\theta \to \phi))$	CP, 12

As usual, following technique (T1), we begin by introducing the given assumptions on lines 1 and 2. Following (T3), we also introduce the assumption $(\psi \to \phi)$ on line 3 with a view to a future use of the EH Rule. Next we turn our attention to the desired conclusion. Following (T7), we introduce the assumption $\forall v\,(\theta \vee \psi)$ on line 4. Since we are now aiming to derive $\neg\forall v\,(\theta \to \phi)$, (T4) suggests that we introduce the assumption $\forall v\,(\theta \to \phi)$ and then seek to derive a contradiction. Since we have the formula $\neg\phi$ on line 1, we shall achieve a contradiction if we can derive ϕ, so we put our trust in (T11) and look for a use of the Tautology Rule. Since the formula

$$((((\psi \to \phi) \mathbin{\&} (\theta \to \phi)) \mathbin{\&} (\theta \vee \psi)) \to \phi)$$

is a tautology, we can derive the formula ϕ by the Tautology Rule. We are then able to follow our intended strategy in five more lines.

We have used the given assumption that the variable v does not occur freely in ϕ on line 12, since without this assumption the use of the EH Rule would not be correct.

7

1	(1)	$\forall v\,\phi$	Ass
1	(2)	ϕ	UE, 1
3	(3)	$\exists v\,\neg\phi$	Ass
4	(4)	$\neg\phi$	Ass
1, 4	(5)	$(\phi \mathbin{\&} \neg\phi)$	Taut, 2, 4
1, 4	(6)	$(\forall v\,\phi \mathbin{\&} \neg\forall v\,\phi)$	Taut, 5
1, 3	(7)	$(\forall v\,\phi \mathbin{\&} \neg\forall v\,\phi)$	EH, 6
1	(8)	$(\exists v\,\neg\phi \to (\forall v\,\phi \mathbin{\&} \neg\forall v\,\phi))$	CP, 7
1	(9)	$\neg\exists v\,\neg\phi$	Taut, 8

The first few lines in this schematic proof are standard. We begin by introducing the given assumption formula, and then apply the UE Rule to it. We aim to derive $\neg\exists v\,\neg\phi$, so we introduce $\exists v\,\neg\phi$ as an assumption and hope to derive a contradiction. Since this assumption begins with an existential quantifier, we also introduce $\neg\phi$ as an assumption with a view to a future use of the EH Rule. It is then very straightforward to obtain the contradiction $(\phi \mathbin{\&} \neg\phi)$ on line 5. But now things become more tricky. This contradiction depends on the assumptions 1 and 4, but we are aiming to derive a contradiction depending on assumptions 1 and 3. In general, the formula $(\phi \mathbin{\&} \neg\phi)$ may contain free occurrences of the variable v, so a use of the EH Rule applied to line 5 would not be legitimate. So we use the contradiction on line 5 to enable us to derive another contradiction on line 6, namely $(\forall v\,\phi \mathbin{\&} \neg\forall v\,\phi)$, in which we can be sure that there are no free occurrences of v. We can then safely apply the EH Rule to line 6 and in this way we are able to complete our proof by contradiction.

Section 2

1

1	(1)	$\forall x \forall y\, x = y$	Ass
1	(2)	$\forall y\, x = y$	UE, 1
	(3)	$x = x$	II
1	(4)	$x = y$	UE, 2
1	(5)	$y = x$	Sub, 3, 4
1	(6)	$x = x$	UE, 2

2 (a)

1	(1)	$x = y$	Ass
	(2)	$(x \cdot x) = (x \cdot x)$	II
1	(3)	$(x \cdot x) = (y \cdot y)$	Sub, 1, 2
	(4)	$(x = y \rightarrow (x \cdot x) = (y \cdot y))$	CP, 3
	(5)	$\forall y\,(x = y \rightarrow (x \cdot x) = (y \cdot y))$	UI, 4
	(6)	$\forall x \forall y\,(x = y \rightarrow (x \cdot x) = (y \cdot y))$	UI, 5

(b)

1	(1)	$x = y$	Ass
	(2)	$(z \cdot x) = (z \cdot x)$	II
1	(3)	$(z \cdot x) = (z \cdot y)$	Sub, 1, 2
	(4)	$(x = y \rightarrow (z \cdot x) = (z \cdot y))$	CP, 3
	(5)	$\forall z\,(x = y \rightarrow (z \cdot x) = (z \cdot y))$	UI, 4
	(6)	$\forall y \forall z\,(x = y \rightarrow (z \cdot x) = (z \cdot y))$	UI, 5
	(7)	$\forall x \forall y \forall z\,(x = y \rightarrow (z \cdot x) = (z \cdot y))$	UI, 6

3 It is sufficient to give examples of formulas ϕ and ψ such that $\phi \vdash \psi$ but the formula $(\phi \rightarrow \psi)$ is not a tautology. It then follows from the Correctness Theorem that ψ is a logical consequence of ϕ, but, as $(\phi \rightarrow \psi)$ is not a tautology, it is not a tautological consequence of ϕ. Example 1.1 provides one such example (there are of course many others): for any formula ϕ, $\exists v\, \neg\phi \vdash \neg\forall v\phi$, but the formula $(\exists v\, \neg\phi \rightarrow \neg\forall v\, \phi)$ is not a tautology.

4 By the Correctness Theorem, it will be enough to show that there is some interpretation of the formal language in which the sentence is false.

(a) In Example 3.7 of *Unit 4* we saw that this sentence is not true in the interpretation given in Example 3.3. As we pointed out there, the non-zero matrix $\begin{pmatrix} 1 & 0 \\ 0 & 0 \end{pmatrix}$ is not the successor of any other matrix in this interpretation.

(b) Take the interpretation with domain the set of integers $\mathbb{Z}$, with $+$ and $\cdot$ interpreted by the usual addition and multiplication on $\mathbb{Z}$, and $'$ interpreted by the function $f\colon \mathbb{Z} \longrightarrow \mathbb{Z}$ given by $f(z) = z^2$. In this interpretation, when x and y are given the values 3 and -3 respectively, the formula $x' = y'$ is true while $x = y$ is false, so that the formula $(x' = y' \rightarrow x = y)$ is false. Thus the formula $\forall x \forall y\,(x' = y' \rightarrow x = y)$ is false in this interpretation.

Any interpretation with domain U which interprets $'$ by a function $f\colon U \longrightarrow U$ which is not one–one will do.

5 Let ϕ be the formula $v = \mathbf{0}$. Then the formula $\exists v\, v = \mathbf{0}$ is true in the standard interpretation as there is indeed a natural number equal to 0. But the formula $\forall v\, v = \mathbf{0}$ is false in this interpretation as there are many natural numbers not equal to 0. So with this choice of ϕ, the formula $(\exists v\, \phi \rightarrow \forall v\phi)$ is false in the standard interpretation. Then by the Correctness Theorem it is not the case that $\vdash (\exists v\, \phi \rightarrow \forall v\phi)$.

There are many other choices of ϕ and interpretations in which $(\exists v\, \phi \rightarrow \forall v\phi)$ is false.

Section 3

1 Suppose that Φ and Ψ are sentences such that $\vdash_T (\Phi \vee \Psi)$, and that neither $\vdash_T \Phi$ nor $\vdash_T \Psi$. Since T is complete, it follows that both $\vdash_T \neg\Phi$ and $\vdash_T \neg\Psi$. The formula $((\neg\Phi \,\&\, \neg\Psi) \rightarrow \neg(\Phi \vee \Psi))$ is a tautology and hence $\neg(\Phi \vee \Psi)$ is a tautological consequence of the formulas $\neg\Phi$ and $\neg\Psi$. Thus $\vdash_T \neg(\Phi \vee \Psi)$. But $\vdash_T (\Phi \vee \Psi)$, and so this contradicts the fact that T is consistent. We deduce that either $\vdash_T \Phi$ or $\vdash_T \Psi$.

2 (a) As $\Phi_1, \Phi_2, \ldots, \Phi_k \vdash \Psi$, repeated use of the CP Rule gives

$$\vdash (\Phi_1 \rightarrow (\Phi_2 \rightarrow (\cdots \rightarrow (\Phi_k \rightarrow \Psi)\cdots)))$$

(b) Suppose that Ψ is derivable from the theory T and that T has as axioms the set of sentences S, so that $T = \mathrm{Th}(S)$. Then Ψ is derivable from assumptions $\Phi_1, \Phi_2, \ldots, \Phi_k$ in T. For $1 \leq i \leq k$, since Φ_i is in $\mathrm{Th}(S)$, it follows that there is a finite set of sentences from S, say Δ_i, such that

$$\Delta_i \vdash \Phi_i$$

Also, from part (a), since $\Phi_1, \Phi_2, \ldots, \Phi_k \vdash \Psi$ we also have that

$$\vdash (\Phi_1 \rightarrow (\Phi_2 \rightarrow (\cdots (\Phi_k \rightarrow \Psi)\cdots)))$$

Thus by writing the formal proofs of $\Delta_i \vdash \Phi_i$ for each i, one under another, followed by the formal proof of $\vdash (\Phi_1 \rightarrow (\Phi_2 \rightarrow (\cdots (\Phi_k \rightarrow \Psi)\cdots)))$, we can obtain a formal proof in which each of the formulas $\Phi_1, \Phi_2, \ldots, \Phi_k$ and $(\Phi_1 \rightarrow (\Phi_2 \rightarrow (\cdots \rightarrow (\Phi_k \rightarrow \Psi)\cdots)))$ occur depending on assumptions only in

$$\Delta_1 \cup \Delta_2 \cup \cdots \cup \Delta_k.$$

Now the formula Ψ is a tautological consequence of the formulas $\Phi_1, \Phi_2, \ldots, \Phi_k$ and $(\Phi_1 \rightarrow (\Phi_2 \rightarrow (\cdots (\Phi_k \rightarrow \Psi)\cdots)))$. Hence, by a single use of the Tautology Rule, we can derive Ψ depending on all the assumptions needed to derive these formulas, that is, on assumptions in the set $\Delta_1 \cup \Delta_2 \cup \cdots \cup \Delta_k$. Since there are only finitely many assumptions in this set, it follows that

$$\Delta_1 \cup \Delta_2 \cup \cdots \cup \Delta_k \vdash \Psi,$$

so that Ψ is in $\mathrm{Th}(S)$, that is, Ψ is in T.

3 We have already seen (in Example 3.7 of *Unit 4* and in the solution to Additional Exercise 4(a) of Section 2 of this unit) that $Q3$ is not true in the interpretation given in Example 3.4 of *Unit 4*. Hence this interpretation is not an interpretation of the theory Q.

In the solution to Problem 3.5(b) of *Unit 4*, we showed that $Q5$ is true in this interpretation. In the solution to Additional Exercise 3(b) for Section 3 of *Unit 4*, we showed that $Q7$ is true in this interpretation. It is easy to check that the remaining axioms of Q, namely $Q1$, $Q2$, $Q4$ and $Q6$, are true in this interpretation.

Thus all the axioms of Q other than $Q3$ are true in this interpretation. Hence $Q3$ cannot be a logical consequence of the other axioms.

In fact it can be shown that no one of the axioms of Q is a logical consequence of the other six axioms. In this sense the axioms of Q are independent of one another.

4 We have seen in Example 3.2 that all the axioms of Q are true in the interpretation $\mathscr{N}^{**}$ given in Example 3.6 of *Unit 4*. Thus by the Correctness Theorem it will be sufficient to show that none of the given sentences is true in this interpretation.

(a) For each element y of the interpretation $\mathscr{N}^{**}$ we have $y + \alpha' = y + \alpha = \beta$, whereas $\alpha'' = \alpha' = \alpha$. Thus for any y there is an element x, namely $x = \alpha$, with $y + x' \neq x''$. Hence the sentence $\exists y \, \forall x \, (y + x') = x''$ is false in $\mathscr{N}^{**}$.

(b) We have $\alpha'' + \beta = \alpha' + \beta = \alpha + \beta = \alpha$, whereas $\beta' + \alpha' = \beta + \alpha = \beta$. Thus $\alpha'' + \beta \neq \beta' + \alpha'$, and hence the sentence $\forall x \, \forall y \, (x'' + y) = (y' + x')$ is false in $\mathscr{N}^{**}$.

(c) For each element y of the interpretation: if $y \neq \alpha$, we have $y \cdot \alpha = \alpha$, and hence $y \cdot \alpha \neq y$; and, if $y = \alpha$, we have $\alpha \cdot 0 = 0 \neq \alpha$. So for each y there is some element x of the interpretation such that $y \cdot x \neq y$. Hence the sentence $\exists y \, \forall x \, (y \bullet x) = y$ is false in $\mathscr{N}^{**}$.

INDEX